Don't Take the Shots

A Critical Assessment of the Ongoing COVID Pandemic

S. H. SHEPHERD

Don't Take the Shots

Cover design: Ljiljana Smilevski.

Contents

Warning: If you ignore the importance of the COVID-19 pandemic and what it portends for the future with further government-sponsored and mandated policies, and neglect to take action that is needed to protect yourself and others, thinking perhaps that things will get better with time, then you are taking a dangerous risk with your life and allowing for a quicker approach of the Great Reset.

Prologue

"If a man writes a book, let him set down only what he knows. I have guesses enough of my own." - Goethe.

This monograph presents facts, not media hype, about the COVID pandemic. Many of the facts, and almost all of the medical advice offered by experienced and well-respected virologists and medical doctors has been ignored and kept from the public, whose very lives are dependent on this information.

The mad rush to develop a vaccine against the virus has resulted in inaccurate guidance provided by government-sponsored medical authorities that has caused many men, women and children to die needlessly due to the SARS-CoV-2 virus. So says the record. Many more people are expected to follow, during this and subsequent years, because of the variants that the virus produces. It has driven physicians and other health care professionals to depart from their Hippocratic oath, which is to serve man over and above all other concerns, including politics and the risk of infection, their departure being to adhere to government-sponsored or mandated policy instead.

This book provides convincing evidence showing why it may not be in everyone's best interest to get vaccinated against COVID-19. It provides the information needed to understand and effectively respond to the pandemic, including ways in which anyone can defend themselves against the virus. It provides the incentives for standing up for the personal freedoms that are guaranteed by the US Constitution but which are now so much under attack.

Preface

The medical authorities and the media claim that the COVID-19 virus was thoroughly researched and studied before the measures to be taken against it were released. But the facts do not support this contention. The medical guidance provided by government-sponsored medical authorities to the medical world and the public did not in 2020, and still does not, adequately address the root cause(s) of the disease, and therefore can be of little use in halting its spread. Furthermore, mandated immunizations are harmful to health and infringe on people's rights.

This book illumines the importance of staying disease free in a world of mandated vaccinations. Having a health issue like high blood pressure, cancer or any other disease puts one at risk of being infected by the SARS-CoV-2 virus or its variants. This book explains why that is so.

It illumines the importance of doing thorough scientific research and investigation before allowing oneself to be vaccinated. It illumines the importance of disseminating proper medical guidance for clinical practice. It relates how medical studies with inconclusive and low-quality data were used to develop government policy that resulted in doing so much harm to so many people.

It reveals many of the mistakes that were made in diagnostic testing which resulted in administering COVID-19 intubation and drugs like Remdesivir to people suspected of having the flu, and then isolating and quarantining them based on established protocol.

Explained is the alarming, parlous nature of the pandemic, the reasons why the studies to isolate and culture the SARS-CoV-2 virus were of poor quality, the very studies that were used to develop the vaccines, how best to remain uninfected, and, if you have already been infected, how to reduce the severity of the disease and the time required for recovery.

If you have not been vaccinated for COVID-19, then don't take the shots. But if you have been vaccinated in order to remain employed, to meet travel requirements or for other reasons, then what you learn in this book will help you in the days and years to come.

Chapter 1 The Ongoing Pandemic

We have been in the COVID-19 virus pandemic since the spring of 2020, and since it is a very contagious virus as evidenced by its quickly spreading throughout the world, all indications are that COVID in its many forms is going to be with us for some time, and perhaps we will never be totally free of it. It is vital, therefore, that a good understanding of the virus be had, and not necessarily one that is based merely on the media narrative. This importance is increased by the fact that age and existing health issues are putting people at greater risk of getting the virus.

The COVID-19 Virus

COVID-19 stands for Coronavirus Disease 2019. "CO" stands for corona, meaning crown, referring to the mace-shaped spike proteins sticking out of the virus, "VI" stands for virus, and "D" stands for disease. The virus that causes COVID-19 is called "SARS-CoV-2," an acronym for "Severe Acute Respiratory Syndrome Coronavirus-2."[1]

Coronaviruses are a family of viruses which consist of many kinds. Most cause mild to moderate upper-respiratory tract illnesses, like the common cold. However, COVID-19 is the third new coronavirus to emerge from animal reservoirs since the beginning of this century to cause serious and widespread illness and death. It was discovered in China in December 2019 and soon spread by travelers to all the industrialized nations of the world. WHO (the World Health Organization) declared it a global pandemic on March 11, 2020.[2] [3]

---

[1] https://www.cdc.gov/coronavirus/2019-ncov/your-health/about-covid-19/basics-covid-19.html.

[2] https://www.niaid.nih.gov/diseases-conditions/coronaviruses.

The exact origin of the COVID-19 virus is unknown, but there appears to be strong evidence implicating it with having been artificially produced through genetic manipulation.[4][5][6]

Like all viruses, COVID-19 consists of nucleic acids (DNA or RNA) surrounded by a protein coat. COVID-19 is an RNA virus with an estimated diameter of between 60 to 140 nanometers. By comparison, a strand of human hair is 400 to 1,000 times larger than the virus. The coat has spikes that facilitate host entry, allowing it to infect a cell. Also like all viruses, the COVID-19 virus is not capable of thriving or reproducing outside of a viable host, which explains why viruses are not considered to be living things by most virologists.[7]

"Viruses are complicated assemblies of molecules, including proteins, nucleic acids, lipids, and carbohydrates, but on their own they can do nothing until they enter a living cell. Without cells, viruses would not be able to multiply."[8]

An important characteristic of viruses is their ability to mutate, and sometimes the mutations result in variants of the virus. At the time of this writing, the latest Omicron subvariants (BA.4.6 and

---

[3] Per the CDC, a pandemic is an outbreak of disease that meets three criteria: 1) illness resulting in death; 2) sustained person-to-person spread; and 3) worldwide spread. (Reference. https://www.ajmc.com/view/a-timeline-of-covid19-developments-in-2020.).

[4] https://economictimes.indiatimes.com/news/defence/view-covid-19-chinas-bioweapon-warfare-strategy-and-global-security/articleshow/83321527.

[5] https://www.heraldopenaccess.us/openaccess/coronavirus-is-a-biological-warfare-weapon.

[6] https://www.youtube.com/watch?v=aF8vx7uutq.

[7] https://www.oralhealthgroup.com/features/the-problems-with-the-covid-19-test-a-necessary-understanding/.

[8] https://www.virology.ws/2004/06/09/are-viruses-living/.

BA.5) are the most contagious so far, accounting for more than 50% of cases.[9] In 2021 it was the Delta variant, which was twice as contagious as the original virus and largely occurring among vaccinated individuals.[10] There is evidence that Omicron variants are the most skillful at out-maneuvering the COVID vaccines.[11] [12]

Meanwhile, medical authorities, such as WHO and Dr. Anthony Fauci, Chief Medical Adviser on COVID-19 for President Biden, emphasize the importance of booster shots and prevention control against COVID-19 such as wearing masks and social distancing, even for those who are already fully vaccinated.[13] We are being told that these measures are the only deterrents against the COVID-19 virus.

Medical Response to COVID-19

When COVID-19 was declared a global pandemic, scientists throughout the world, including special teams consisting of physicians and epidemiologists, were mobilized to study the virus.[14] On June 16, 2020, it was announced by officials with the United States' Operation Warp Speed, a project to rapidly develop and deploy a COVID-19 vaccine, that the vaccine would be offered

---

[9] https://www.yalemedicine.org/news/5-things-to-know-omicron.

[10] https://www.cdc.gov/coronavirus/2019-ncov/variants/delta-variant.html?s_cid=11516:covid%20vaccine%20and%20delta%20variant:sem.ga:p:RG:GM:gen:PTN:FY21.

[11] https://www.cdc.gov/coronavirus/2019-ncov/variants/about-variants.html?CDC_AA_refVal=https%3A%2F%2Fwww.cdc.gov%2Fcoronavirus%2F2019-ncov%2Fvariants%2Fvariant.html.

[12] https://www.nbcnews.com/think/opinion/omicron-proves-covid-vaccine-working-breakthrough-cases-misleading-term-ncna1286730.

[13] https://www.nbcboston.com/news/local/what-are-the-side-effects-of-the-pfizer-moderna-covid-booster-shots/2584210/.

[14] https://www.ncbi.nlm.nih.gov/pmc/articles/PMC7166628/.

free of charge to elderly patients and to populations who cannot afford it.[15]

The CDC- and FDA-approved PCR (Polymerase Chain Reaction) test, technically known as the LabCorp COVID-19 RT-PCR, is a molecular diagnostic test to determine if a person is infected with the COVID-19 virus. It analyzes a sample taken from the person to see if it contains genetic material from the virus.[16]

While it is not the purpose of this discussion to find fault with medical researchers and authorities, the urgent need to develop a vaccine and testing program for COVID-19 resulted in some serious mistakes made by the medical authorities.

Although a vaccine for the COVID-19 virus has been produced and continues to be in use throughout the US and the world, with more vaccines in the offing for the variants, it was revealed in a CDC report issued on July 13, 2021[17] that neither the FDA nor the CDC had a trace of the virus, known as a virus isolate, for the diagnostic testing program, the very program used to determine if a person has the virus.

A virus isolate means a virus sample that has been isolated from an infected host.

The cited July 13, 2021 report stated:

"Since no quantified virus isolates of the 2019-nCoV were available for CDC use at the time the test was developed and this study conducted, assays [diagnostic tests] designed for detection

---

[15] https://www.ajmc.com › view › us-releases-more-detail.

[16] https://www.webmd.com/lung/coronavirus-glossary#1.

[17] https://www.naturalnews.com/2021-07-30-cdc-fda-faked-covid-testing-protocol-by-using-human-cells-mixed-with-common-cold-virus.html.

of the 2019-nCoV RNA were tested with characterized stocks of in vitro transcribed full-length RNA (N gene; GenBank accession: MN908947.2)."[18] [19]

In other words, a COVID virus isolate was not available for developing and calibrating the PCR test that was approved for use on July 24, 2020, approximately a year before the cited 2021 CDC disclosure report came out. Instead, they used human cells and RNA fragments from a common cold virus.

Since the PCR test was developed without actual COVID samples, it could not differentiate between the common cold and COVID. This resulted, during nearly all of 2020, in the number of influenza cases nearly disappearing, as influenza was re-labeled "COVID" due to the faulty testing.[20]

"In essence, the medical establishment simply took all the people who would normally be diagnosed with colds and the flu, and shifted them into the "COVID" category in order to push a COVID mass hysteria narrative that would drive people into vaccines. The vaccines, then, were formulated with spike protein toxic nanoparticles to cause the "delta" panic wave, which is largely occurring among vaccinated individuals."[21]

When developing deterrents for infectious diseases, it is not wise to proceed at warp speed.
The following four paragraphs are excerpted from the footnoted

---

[18] Ibid.

[19] https://www.reuters.com/article/uk-factcheck-cdc/fact-check-this-cdc-document-does-not-say-that-that-sars-cov-2-doesnt-exist-idUSKBN27633R.

[20] https://www.naturalnews.com/2021-07-30-cdc-fda-faked-covid-testing-protocol-by-using-human-cells-mixed-with-common-cold-virus.html.

[21] Ibid.

website dated February 8, 2021.[22]

"Determining the accuracy and reliability of a test for a pathogen requires the presence of a gold standard. A gold standard provides authoritative, and presumably indisputable, evidence that a condition does or does not exist.

"Developing a gold standard for SARS-CoV-2 [COVID-19] requires isolating the whole virus, not just fragments of it, and showing that the isolated virus is capable of reproducing itself in culture cells.

"A recent extensive review concluded that studies to isolate and culture SARS-CoV-2 were of poor quality and lacked assessment against an acceptable gold standard. Another report noted that while in previous international health emergencies, viral isolates were available to validate tests, in the case of SARS-CoV-2, virus isolates or samples from infected patients have so far not become available to international communities.

"Indeed, in July 2020, the CDC stated that no quantified virus isolates for the SARS-CoV-2 existed."

The worst part is that a large number of people – the exact number cannot be determined – were identified as having the COVID virus when they really had the flu.

A NIH US National Library of Medicine article published on August 31, 2020 and entitled, "Laboratory Diagnosis of COVID-19," which had the objective of reviewing the literature on laboratory diagnostic testing of COVID-19, concluded the following: "The identification of genetic material of the virus by RT-PCR is the gold standard test, but its sensitivity is not satisfactory. The

---

[22] https://www.oralhealthgroup.com/features/the-problems-with-the-covid-19-test-a-necessary-understanding/.

diagnosis of COVID-19 should be based on clinical data, epidemiological history, tests for etiological diagnosis, and tests to support the diagnosis of the disease and/or its complications. New diagnostic methods with higher sensitivity and specificity, as well as faster results, are necessary."[23]

Despite these facts, or in spite of them, many of the NIH, CDC and other medical authority reports that I have seen in my research on this subject dated after July 2020 appear to be defensive statements made to pacify the American public that everything is being done properly and that the PCR test being used to determine if a person has COVID-19 can be relied on as accurate, reliable and a gold standard.

In addition, similar articles published throughout 2020 appear in some way to be misleading, or are not sufficiently detailed to tell whether or not a virus isolate of COVID-19 was, in fact, used for calibrating and testing purposes.

An example is the article published by the CDC dated December 29, 2020, which states the following. [24]

"SARS-CoV-2, the virus that causes COVID-19, was isolated in the laboratory and is available for research by the scientific and medical community.

"On January 20, 2020, CDC received a clinical specimen collected from the first reported U.S. patient infected with SARS-CoV-2. CDC immediately placed the specimen into cell culture to grow a sufficient amount of virus for study. On February 2, 2020, CDC generated enough SARS-CoV-2 grown in cell culture to distribute

---

[23] https://www.ncbi.nlm.nih.gov/pmc/articles/PMC7456621/.
[24] https://www.cdc.gov/coronavirus/2019-ncov/lab/grows-virus-cell-culture.html.

to medical and scientific researchers. On February 4, 2020, CDC shipped SARS-CoV-2 to the BEI Resources Repository. An article discussing the isolation and characterization of this virus specimen is available in Emerging Infectious Diseases."

Please note that the last two sentences in the above excerpt are key to the interpretation of the article, because upon them rests the validity of the report. The article referenced in the last sentence, which is dated June 2020,[25] includes the following Abstract.

"The etiologic agent of an outbreak of pneumonia in Wuhan, China, was identified as severe acute respiratory syndrome coronavirus 2 in January 2020. A patient in the United States was given a diagnosis of infection with this virus by the state of Washington and the US Centers for Disease Control and Prevention on January 20, 2020. We isolated virus from nasopharyngeal and oropharyngeal specimens from this patient and characterized the viral sequence, replication properties, and cell culture tropism. We found that the virus replicates to high titer in Vero-CCL81 cells and Vero E6 cells in the absence of trypsin. We also deposited the virus into 2 virus repositories, making it broadly available to the public health and research communities. We hope that open access to this reagent will expedite development of medical countermeasures."

Please note the first two sentences in the above excerpt. They imply that the patient from which the virus was isolated was correctly diagnosed as having the COVID-19 virus. However, there is a huge gap of factual information missing between the first and second sentences which begs the question, was a COVID-19 virus

---

[25] https://pubmed.ncbi.nlm.nih.gov/32160149/.

15

isolate available {for example, from China} when the patient was diagnosed on which to base the conclusion that the patient had the COVID-19 virus? In other words, did the state of Washington and the CDC have a gold standard upon which to base their diagnosis?

The answer appears to be an emphatic "No," as indicated by the CDC report previously cited dated July 13, 2021, the day that the material allegation of the facts held by CDC was disclosed. No gold standard test was available upon which to base a COVID-19 diagnosis, at least in this country, up to at least mid-2021 (but even that is now dubious). This is substantiated by the article previously cited dated February 8, 2021. The question remains, unfortunately unanswered, whether a verifiable sample of COVID-19 was actually cultured *after* mid-2021.

Most people are accepting the media narrative without question. Nevertheless, other difficulties remain with the diagnostic testing for COVID-19.

Test Criteria to be Satisfied

Certain criteria have been established by virologists and microbiologists that should be satisfied to determine a microbe (either bacteria or virus) and a pathogen causal relationship. They were first established by the German physician, Robert Koch, in 1890. Known as Koch's postulates, they became a scientific standard for evidence that established the credibility of microbes being the cause of pathogens, which led to the development of modern microbiology.

Koch's postulates have since evolved to accommodate the discovery and improved understanding of viruses and their difficulty to be purified and cultured. In 1996 a set of molecular guidelines were proposed by Fredricks and Relman of the Stanford University School of Medicine to be satisfied before a causal (not casual) relationship can be said to exist between a virus and an infection. These seven guidelines have subsequently been accepted by most practitioners in the field. [26] [27] [28]

However, for COVID-19, such a causal relationship was, at least until April 2020, not established.[29] An article published in the Journal of Medical Virology after the virus was discovered in Wuhan, China stated:[30]

"The data collected so far is not enough to confirm the causal relationship between the new type of coronavirus and the respiratory diseases based on classical Koch's postulates or modified ones as suggested by Fredricks and Relman."

Nevertheless, it has been accepted that SARS-CoV-2 is the etiologic agent of COVID-19.[31]

Other difficulties exist as well. The following discussion provides some of the details of COVID-19 testing that make for a valid diagnostic virus test.

Details of the COVID Test

The COVID-19 lab test (Polymerase Chain Reaction, PCR) looks

[26] https://en.wikipedia.org/wiki/Koch%27s_postulates.
[27] https://www.oralhealthgroup.com/features/the-problems-with-the-covid-19-.
[28] www. reasearchgate.net > publication > 14534757 F.
[29] https://www.oralhealthgroup.com/features/the-problems-with-the-covid-19-.
[30] https://pubmed.ncbi.nlm.nih.gov/31950516/.
[31] https://www.oralhealthgroup.com/features/the-problems-with-the-covid-19-test-a-necessary-understanding/.

for the genetic material of the SARS-CoV-2 virus. The genetic blueprint for SARS-CoV-2 is coded in RNA not DNA.[32] The PCR test, which was developed over 35 years ago and is one of the most widely used lab tests for determining if the host has viruses, is highly sensitive to the presence of RNA. This means that it can detect any RNA present which might, or might not, be part of the SARS-CoV-2 virus. [33]

The test requires a sample be taken from a person by a health care provider. A swab is inserted deep into a person's nostril, but other samples can also be taken from a throat or saliva swab.

PCR testing begins by converting the virus's RNA into DNA because DNA is a lot more stable than RNA. The PCR machine, which involves heating, cooling and heating the sample again and again, makes millions of copies of the DNA by running multiple "cycles." The primary function of the test is to make many copies of a specific region of DNA such that this target area can be better analyzed.[34] [35]

"This process is called amplification and is extremely important in finding even the smallest amounts of DNA. As more cycles are run, more copies of the DNA are made – doubling every time it is copied – and making it easier to find. If the piece of DNA cannot

---

[32] DNA (deoxyribonucleic acid) and RNA (ribonucleic acid) and RNA are genetic material found in living things.
[33] https://www.publichealthontario.ca/en/about/blog/2021/explained-covid19-pcr-testing-and-cycle-thresholds.
[34] https://www.publichealthontario.ca/en/about/blog/2021/explained-covid19-pcr-testing-and-cycle-thresholds.
[35] https://www.oralhealthgroup.com/features/the-problems-with-the-covid-19-test-a-necessary-understanding/.

be copied, there is no virus in the sample, or there is such a low amount that even the test cannot detect it."[36]

Ontario, Canada uses a cut off or stopping point, also called a threshold, of 38 cycles for positive results of the test. This means that if the virus is found at or before 38 cycles are completed, then the test is considered positive. For negative results, it uses a cutoff point of 40 cycles.

The number of cycles used for a positive and negative test are not standardized but varies within provinces or countries, which adds to the unreliability of the COVID-19 test.[37]

"Recent [2021] papers have suggested that greater than 24 cycles should not be used to infer the presence of a "live or infectious" virus since above that level the exquisite sensitivity of the test will amplify sequences of viruses from other sources, and these sources should be dead, or non-infectious, SARS-CoV-2, or general cell debris, endemic coronaviruses, other pathogens, and contamination during collection, transportation and preparation of samples."[38]

What it means is that some people testing positive for COVID-19 may not be infected by it at all, and do not have to undergo the stringent isolation and retesting requirements.

In a September, 20, 2020 Web article entitled, "International Experts Suggest That Up to 90% COVID Cases Could be False

---

[36] https://www.publichealthontario.ca/en/about/blog/2021/explained-covid19-pcr-testing-and-cycle-thresholds.

[37] https://www.oralhealthgroup.com/features/the-problems-with-the-covid-19-test-a-necessary-understanding/.

[38] https://journals.plos.org/plosone/article?id=10.1371/journal.pone.0256835.

Positives," Dr. Barbara Yaffe, Director of Communicable Disease Control, Toronto Public Health, told the media:

"In fact, if you are testing in a population that doesn't have very much COVID, you'll get false positives almost half the time. That is, the person actually doesn't have COVID, they have something else, or they may have nothing."[39]

As stated in a May, 2021 article published in Lancet:

"COVID-19 has had negative repercussions on the entire global population. Despite there being a common goal that should have unified resources and efforts, there have been an overwhelmingly large number of clinical trials that have been registered that are of questionable methodological quality." - Jay J H Park, MSc., Department of Experimental Medicine, University of British Columbia, Vancouver, BC, Canada.

A recent (1/7/22) website article announced that nearly half of all persons in Ontario, Canada, have refused to take the brand name Moderna Spikevax COVID-19 vaccine.[40]

A Nebraska Medicine Web article published August 19, 2021, as well as other Web articles of a similar nature and date, reported that the CDC has recalled the original PCR test and replaced it with a test called the multiplexed method that tests for both COVID and influenza at the same time.[41]

---

[39] https://westphaliantimes.com/international-experts-suggest-that-up-to-90-of-canadian-covid-cases-could-be-false-positives/.

[40] https://globalnews.ca/news/8495575/moderna-covid-shot-refusal-ontario-omicron/

[41] https://www.nebraskamed.com/COVID/pcr-test-recall-can-the-test-tell-the-difference-between-covid-19-and-the-flu.

But what I consider is the amazing part of the article is that it claims that the standard PCR test was "extraordinarily accurate" in identifying the right virus (SARS-CoV-2), which contradicts the July 13, 2021 report cited above which stated that since no quantified virus isolates of the 2019-nCoV were available for CDC use at the time the test was developed, the standard PCR test could not differentiate between influenza and COVID-19. What Mark Twain is credited to have said appears to be true:

"It is easier to fool people than to convince them that they have been fooled." - Mark Twain.

It would appear that the tact taken by the CDC currently is to keep in the background its former failures in connection with the pandemic and its admittance that the standard PCR test was fraudulent. While I believe that this stance or tact makes sense in that it keeps many people from taking legal or other action against the CDC, one would suspect that its primary purpose is to save face and encourage positive American opinion about these tests.

For all the above reasons, it behooves every one of us to be skeptical of the many claims being made by our medical authorities regarding the safety, efficacy and reliability of the COVID-19 testing, and the vaccines that have been produced. The facts that have been presented characterize the failure by the governments of the world to appreciate the limitations of the COVID-19 tests on which all policies and precautions to eliminate or reduce the spread of the disease are based.

VEKLURY (Remdesivir)

A PCR test is required for whoever is admitted in hospital or clinic.

If positive, they are treated for COVID-19 according to a protocol, Remdesivir is administered by intravenous infusion (slow injection), and the person may be subjected to mandatory quarantine, which is the same as being forcibly incubated.[42]

Randomized clinical trials using this drug versus a placebo revealed that 54% of patients were more likely to have improved clinical status at day 15 than those taking the placebo. Our medical community is calling this a reason for using it. It is interesting to note that those who received remdesivir had a median recovery time of 10 days compared to 15 days recovery time among those who received the placebo.[43 44] This would indicate that the drug can hasten but is not in itself necessary for recovery from COVID-19.

Nevertheless, there is strong evidence suggesting that remdesivir causes more harm than good since it has some serious side effects, among which are hypersensitivity, including infusion-related and anaphylactic reactions which can be severe, life-threatening allergic reactions. and transaminase elevations, a term generally applied to inflammation of the liver.[45 46]

A PLOS ONE journal article published on September 1, 2021 revealed that the net impact (direct and indirect effects) of the COVID-19 pandemic on the US population in 2020 was 375,235 excess deaths, with 83% attributable to direct, and 17% attributable to indirect effects of the virus.[47]

---

[42] It was announced by the CDC on August 17, 2022 that this requirement may be rescinded.

[43] https://www.vekluryhcp.com/efficacy/?

[44] https://www.nejm.org/doi/full/10.1056/nejmoa2007764.

[45] https://www.vekluryhop.com/important-safety information/?gclid= Cj0KCQjwqp-.

[46] https://www.mayoclinic.org › syc-20351468.

[47] https://journals.plos.org/plosone/article?id=10.1371/journal.pone.0256835.

In a scientific report published February 18, 2021, the years of life lost across 81 countries due to COVID-19 attributable deaths were over 20.5 million years.[48]

Cost to Patients

As time goes on, it seems that the COVID-19 situation is getting worse day by day. For example, at the time of this writing, the cost to patients of COVID-19 medical care is skyrocketing, and one who happens to be in the insurance business may say, "at last!" What used to be free tests and professional care in 2020 are now being billed to the patients, and this will no doubt continue as we get further into the pandemic.

A New York Times article by Sarah Kliff, published on September 2, 2021 and updated September 9, 2021 stated that for each COVID-19 intubation administered by hospital staff, the hospital receives $40,000, a cost that is being passed on to many patients.[49]

As a result of the COVID-19 vaccine having been given to hundreds of millions of people around the world, it has become the most profitable vaccine in history. Global sales of the vaccine are expected to top $50 billion in 2021 alone.[50]

Growing Fears

A growing fear of the effects of the COVID-19 vaccine is what many are calling Draconian requirements. In September of 2021, President Joe Biden signed an executive order requiring that all workers in the Executive Branch be vaccinated against COVID-19. He also signed an order to require businesses with more than 100

---

[48] https://www.nature.com/articles/s41598-021-83040-3.
[49] https://www.nytimes.com/2021/09/02/upshot/covid-medical-bills.html.
[50] https://www.nature.com/articles/d41586-021-02483-w.

workers to mandate coronavirus vaccinations.[51] Many states complied with the mandate, and similar mandates are likely in the days to come unless people wake up and stand up for their rights instead of handing them away. Meanwhile, it appears that our mainstream medical institutions are becoming glorified drug dealers and minions of the government- controlled medical system, with very little capacity for critical thought.

But perhaps the worst fear of all today in connection with the COVID-19 vaccine is that it, and the subsequently-required booster shots against variants of the virus, have the propensity to cause, as they have in animal tests with messenger RNA vaccines, slow death and reduced life expectancy. The animals in the tests died of antibody dependent enhancement (ADE). They didn't die right away but months later when they were reintroduced to the virus. The virus triggered a cyclone storm of runaway inflammatory response in the animals which proved fatal.

"ADE occurs when the antibodies generated during an immune response recognize and bind to a pathogen, but they are unable to prevent infection. Instead, these antibodies act as a "Trojan horse," allowing the pathogen to get into cells and exacerbate the immune response."[52]

It appears that ADE is a potential risk for anyone who has been vaccinated against the SARS-CoV-2 virus. When we ponder these things, as well as the fact that the third leading cause of death in the US is the inappropriate application of pharmaceutical and

---

[51] https://www.whitehouse.gov/briefing-room/presidential-actions/2021/09/09/executive-order-on-requiring-coronavirus-disease-2019-vaccination-for-federal-employees/.

[52] https://www.chop.edu/centers-programs/vaccine-education-center/vaccine-safety/antibody-dependent-enhancement-and-vaccines.

medical interventions,[53] [54] it is a frightful world, indeed, that we are passing on to our progeny.

All of this discussion underlines the vital importance that every person understand how to maintain their own health so they can decrease their chances of ending up in the hospital. It is important to remember that the basic causes of practically all diseases are bodily acidity and toxemia. Also, all illnesses are an attempt by the body to cleanse itself of accumulated toxins. Symptoms like rashes, coughing and fever are the way the body normalizes things and should never be suppressed. Mainstream medicine, however, is concerned only with suppressing symptoms at great cost to the patient's health in the long run.

Two links that describe in more detail the dangers of COVID testing and vaccination and how individuals of gender and ethnic group in this country are affected by the vaccination, are given below. The first is a longer interview but is extremely informative. Those who are interested will want to watch these videos.

https://ugetube.com/watch/dr-ardis-which-is-more-deadly-covid-or-the-hospitals-flyover-conservatives_RZVejBvV1dHcR7j.html#

https://renz-law.com/media-news-feed/.

Dr. Sherri Tenpenny in her 2021 downloadable report entitled "20 Mechanisms of Injuries (MOI),[55] identified 20 mechanisms of injuries, 20 ways COVID-19 injections can make you sick or even kill you. For example, they can cause death through anaphylactic shock or anti-Inflammatory macrophages. I highly recommend

---

[53] https://pubmed.ncbi.nlm.nih.gov/25355584/.

[54] https://www.vekluryhop.com/important-safety-information/?gclid=Cj0KCQjwqhttps://www.hopkinsmedicine.org/news/medicine.org/news/media/releases/study_suggests.

[55] Available on the Web for download.

that anyone even considering taking the government mandated injections read this report.

Masks

At the time of this writing, almost everyone in the US has given up wearing some kind of mask in public, but it was mandatory for the first two years of the pandemic.

The surgical mask, the mask considered to be the best to wear against COVID-19 and other viruses, is the mask the government helped produce in 2020 under the Trump administration. It is called the N95 mask. The mask is designed to block large-particle droplets, splashes, sprays, or splatter that may contain germs (viruses and bacteria)."[56]

While the N95 mask may be effective in blocking splashes or large-particle droplets, a face mask of any kind cannot block or filter out very small particles in the air that may be transmitted by coughs, sneezes, or certain medical procedures. The N95 mask, like all other masks worn against the coronavirus infection, while preventing someone's drool or spittle, or sneeze particles, from reaching your mouth or nose, cannot prevent germs or viruses from reaching you.

If a person comes in close contact with someone who is infected with the COVID-19 virus, the best of masks cannot prevent the virus from being transmitted to them, for viruses are the smallest of microbes.

"While a surgical mask may be effective in blocking splashes and

---

[56] https://www.fda.gov/medical-devices/personal-protective-equipment-infection-control/n95-respirators-surgical-masks-and-face-masks.

large-particle droplets, a face mask, by design, does not filter or block very small particles in the air that may be transmitted by coughs, sneezes, or certain medical procedures."[57]

The Best Way to Fight Off COVID-19

The vast literature on the effects of foods and nutrition on human health of which I am familiar after years of studying and, just as importantly, putting into practice the principles of optimum health, indicates most emphatically that a strong immune system can not only fight off viruses that originate from outside the body and prevent them from taking root in the human soil of the body, but also, if a person is already infected, can greatly reduce its severity and the time in which a person remains infected. This has been confirmed countless of times by people of all walks of life, age, gender, ethnic group and nationality.

"Many people have a strong and healthy immune response to the coronavirus which may give them protection against future infections."[58]

It only makes sense. 70-80% of the cells of the immune system reside in the gut.[59] The gut microbiota aid in food digestion and fight off invading microorganisms. A nutrient rich, well-balanced diet provides the beneficial bacteria in your intestines with plenty of nutrients to thrive on.[60] Certain diseases, like cancer ,heart disease and kidney disease repress the immune system which allows increased attacks by unfriendly microbes, including the

[57] Ibid. (https://www.fda.gov/medical-devices/personal-protective-equipment-infection-control/n95-respirators-surgical-masks-and-face-masks.)
[58] https://www.immunology.org/news/coronavirus-immunology-qa-what-you-need-know-about-our-new-report.
[59] https://pubmed.ncbi.nlm.nih.gov/33803407/.
[60] https://www.novanthealth.org/healthy-headlines/do-you-need-probiotics.

smallest of microbes, the viruses. This is a well-proven medical fact. But a strong immune system can prevent unfriendly microbes like COVID 19 from spreading through the body. This wealth of knowledge has not been disseminated as well as it should in media sources like the Web/Internet. In fact, some of websites claim that there is no immunity to COVID-19. Nevertheless, this vital information is available to all by doing research and spending the time to understand the somewhat technical information that is gathered.

One may wonder, as I have, why we do not all get diseases, such as polio, diphtheria, typhoid or cholera. The answer from a nutritionist's viewpoint is that there is no suitable soil in our bodies in which these diseases can take root. Based on my research and as discussed in this book, there appears to be the only reason we do not all get diseases. Microorganisms cannot harm us as long as our organism functions in a normal, healthy way. Microbes such as viruses are harmless if the body's defensive mechanisms, including the immune system, are in good working order. A weakened immune system can be caused by drugs, improper diet, unhealthy lifestyles, health issues such as obesity, advanced age, diabetes, lung disease and heart disease, and these things can put people at increased risk for COVID-19 and its variants.[61] [62] [63]

I have never taken the shots and I have never been in better health. I leave it to the naysayers to make little of such a seeming contradiction.

---

[61] https://www.hopkinsmedicine.org/health/conditions-and-diseases/coronavirus/coronavirus-the-covid19-vaccine-and-epilepsy.

[62] https://www.cdc.gov/coronavirus/2019-ncov/need-extra-precautions/people-with-medical-conditions.html.

[63] https://katv.com › news › local › high-blood-pressure-d...

Anyone who eats whole plant foods almost exclusively will have all the enzymes, nutrients and antioxidants the body needs to support a strong immune system, and need not fear getting infectious diseases like COVID-19. In other words, to ensure your immune system protects you from infectious organisms, take care of your gut microbiome by eating nutritious foods, and stay away from foods or their combinations that cause diarrhea, gas and constipation which decrease the strength of the immune system.

According to an August 11, 2021 Web article,[64] the people with the highest risk of getting COVID-19 are those 85 and older. That also is logical since, as explained in the article, those of great age have less effective immune reactions to the virus and are more likely to suffer from high blood pressure and other health issues which weaken the immune system. In addition, as one ages, the tissues of the lungs deteriorate, making it harder for them to deal with a COVID infection.

It is for this reason that the rest of the book focuses on human diseases and their relationship to foods that are eaten.

I have never seen, but would like to see, the statistics on those who have contracted the COVID virus and who have adopted a whole plant food diet and having little, if anything, to do a meat and dairy based diet, which has been proven beyond anyone's skepticism to damage health, including lowering the strength of the immune system. Assuming a correct diagnosis and not one that is really the flu, I would venture to guess that the statistics would be very low indeed.

"The adulterated, unnatural, false, man-made foods of present-day civilization are the underlying physiological causes of all evils

---

[64] https://www.webmd.com/lung/whos-at-risk-covid-19#1.

which humans are prey – especially of all kinds of diseases. Health will not return, nor can it be regained through drug remedies or the various medical treatments, since supreme, absolute, paradisiacal health is ruled by the laws of diet!" - Professor Arnold Ehret, *Thus Speaketh the Stomach and the Tragedy of Nutrition*.

The upshot is that we do not have to be tragic victims of infectious or non-infectious diseases. We can do something about it.

I've included the following to round out this discussion and lighten any misplaced blame attributed to the testing errors that have been described.

Of the Diseases this Year

"This year the stone-blind shall see but very little; the deaf shall hear but poorly; and the dumb shan't speak very plain. Whole flocks, herds, and droves of sheep, swine and oxen, cocks and hens, ducks and drakes, geese and ganders shall go to the pot, but the mortality will not be altogether so great among cats, dogs and horses. As for old age, it will be incurable this year because of the years past. And towards the fall some people will be seized with an unaccountable inclination to roast and eat their own ears. Should this be called madness, doctors? I think not. But the worst disease of all will be a certain most horrid, dreadful, malignant, catching, perverse and odious malady, almost epidemical insomuch that many shall run mad upon it. I quake for very fear when I think on it, for I assure you very few will escape the disease, which is called by the learned Albumazar, Lacko'mony." *- Poor Richard's Almanac: 1739.*

Chapter 2 The Causes of Disease

"The doctor of the future will give no medicines, but will interest his patients in the care of the human frame, in diet, and in the causes and prevention of disease." - Thomas Edison.

Are not our lives more painful and shorter than they should be? Health issues uncommon in the past are frequently experienced today. Widespread and serious illnesses and diseases strike everyone, everywhere. How many people do you know who are suffering from a health affliction or disorder? How many readers of this book have heart disease or some other health issue they're wrestling with?

Diseases seem to wait in ambush, ready to strike and slay. But what are they really, at least according to our best scientific understanding, and what does this understanding tell us about how diseases may be cured? This chapter explores these topics.

There are, no doubt, hundreds of diseases that afflict humankind, including all the fevers, tumors and contagion of human life. It is not the purpose of this book to chronicle all human diseases or to give an account of the known causes of them, but to explore some of the attempts made by man to understand, control and cure the diseases that beset him. Two views of diseases will be presented, the view commonly held by the mainstream medical profession, and the view held by modern nutritionists.

Bear in mind that since human diseases appear to have certain causes, and we have evidence that there are methods by which they can be avoided, there must be ways to cure them.

Mainstream Medicine View

The deadliest diseases in this country, and the most feared, are heart disease; cancer of the organs, including colon, bladder, lung and breast cancer, kidney disease, COVID-19, and the neurological diseases, including Alzheimer's disease and Parkinson's disease. According to mainstream medicine, the causes of these diseases are primarily foreign substances, including but not limited to bacteria, viruses and poisons that get into, or invade, the body.

The theory of infection is built around microorganisms; in short, microbiology, a product of laboratory work based on the proposition that microorganisms, which include bacteria and viruses, are responsible for the majority of our diseases. The attempts made by mainstream medicine to cure diseases are primarily efforts that are aimed at eradicating the foreign substances and organisms that are believed to cause the diseases. Of course, vaccines may also be used.

The methods used by mainstream medicine to treat diseases include pharmaceutical drugs, antibiotics such as penicillin and erythromycin, and medical intervention such as hospitalization and surgery. Many of the drugs must be improved yearly by medical laboratories to counteract the proliferation of drug-resistant organisms that result from their use. Chemotherapy, radiation therapy and vaccines are used. These methods are, for the most part, all that mainstream medicine has in its arsenal against diseases.

Standard medical treatments focus on achieving health through drugs. For headaches, take aspirin or a similar product; for stomachaches or heartburn, take antacids; for infections, take antibiotics; for constipation, take laxatives; for cancer, take chemotherapy or radiation therapy. But even considering drugs like penicillin which can halt the spread of a disease, drugs do not possess the ability to cure, and there are risks associated with all drugs. Also, drugs have been known to lodge in the system for decades after their use.[65]

Diseases which are today considered eradicated, such as malaria, smallpox and poliomyelitis, are not eradicated in the true sense of the word, because the viruses or parasites associated with them still exist. They are merely prevented from causing excessive harm through vaccination, which is the most common form of immunization used against diseases.

There are, of course, exceptions. Leprosy and inflammation of the eyes (uveitis, a disease that can cause blindness), can both be cured through drug therapy as long as sanitary measures are also put in place, unless, in the case of eye inflammation, the cause is an autoimmune disorder. Both diseases were the scourge of the Middle East 200 years ago when many of the population suffered terribly from them. Physical blindness today is often prevented merely through sanitary measures.

Permanent cures of diseases by mainstream medicine are rare.[66] It is believed that many of the cures that are attributed to mainstream medical treatments would have occurred naturally without them.

---

[65] From the book, Prof. Arnold Ehret's Mucusless Diet Healing System: Annotated, Revised, and Edited by Prof. Spira.

[66] For example, the average cure rate for all types of cancer by mainstream medicine, except for skin cancer, is 17% per G. Edmond Griffin's book, World Without Cancer. It is believed to be more like 6% per Rich Anderson's book, Cleanse & Purify Thyself. A 2019 American Cancer Society Web article states that most cancers cannot be cured, but some can be controlled for months or even years. A cancer.net article sponsored by the American Society of Clinical Oncology (ASCO) states that chronic cancer is cancer that cannot be cured.

Viruses and bacteria are everywhere and exist in everyone, but before they can proliferate and thrive in the body, they must have suitable soil. Nutritionists believe that it is the waste material in the body that affords the germs the suitable soil for them to proliferate and produce the symptoms of disease.

Many nutritionists, including those who are referenced in this book, contend that almost all human diseases are the result of internal poisoning, typically organ or blood poisoning, that occurs when the tissues of the body become congested with toxic waste material due to constipation. To put it another way, they believe that disease is the result of intestinal occlusion, the clogging up of food wastes in the large intestine (colon), and that correcting the situation through proper diet so that the number of defecations per day is about the same as the number of meals eaten per day is the only way the body can rid itself of disease, recover and then resume a life of health.

If the natural self-healing process is repressed through medication (drugs), and we continue the consumption of foods and drinks that are harmful to the body, then the cure is thwarted.

They believe that disease symptoms are manifested as a result of this poisoning, and that they are signs the body gives when it attempts to eliminate the poisons. Malaise and fatigue are two typical signs of internal poisoning, and so are sores and rashes. In other words, disease is an attempt to rid the organism of the foul material with which it is contaminated. Also, disease symptoms are the outward manifestation of an automatic internal bodily cleansing process. For example, nausea and fever result from the body's natural attempts to rid itself of morbid, toxic material that has accumulated in the tissues. They may be triggered by, but they are not caused by, "infectious organisms" that invade the body.

What do nutritionists say is the answer?

"In food lies 99.99% of the causes of all disease and imperfect health of any kind. Consequently, all healing, all therapeutics will continue to fail as long as they refuse to place the most important stress on diet." - Arnold Ehret, *Mucusless Diet Healing System*.

The cure for disease, then, is not to treat the symptoms, including the pains that may result from the disease, but the underlying causes of the symptoms.

"The foods you consume can heal you faster and more profoundly than the most expensive prescription drugs, and more dramatically than the most extreme surgical interventions, with only positive side effects. They can prevent cancer, heart disease, Type 2 diabetes, stroke, macular degeneration, migraines, erectile dysfunction, and arthritis – and that's only the short list." - T. Colin Campbell and Howard Jacobson, *Whole, Rethinking the Science of Nutrition*.

Many of the diseases that afflict us appear to be the result of either an unwitting ignorance of the unalterable laws of nature, or a willful refusal to abide by these laws. Adopting a whole plant food diet and restricting food intake to manageable amounts of the most easily digestible foods helps to ensure that every seed of sickness is extirpated from the body.

"Man's health or his disease of every description, directly result from food intake. His state of mind may be a contributing factor, but the fall of mankind in the final analysis is "sin of diet." The real physiological cause of all evils, especially the physical ailments of mankind can be traced directly to the present day accepted diet of civilization." - Arnold Ehret, *Rational Fasting and Roads to Health and Happiness*.

It has been my experience and observation that we are more than capable of causing our own sicknesses. We unwittingly bring sickness and disease upon ourselves through dietary practices that are contrary to the laws of nature. The body requires a variety of nutrients to keep itself in health, but seldom finds them in the foods that are commonly eaten. It requires enzymes for proper digestion, but they are destroyed by heat, and almost all refined and processed foods are heat treated. Deprived of the nutrients and other essentials it needs to enhance the body's self- cleansing process, the natural law of cause and effect determines the outcome, which is ill-health.

An old saying is:

One quarter of what you eat keeps you alive. The other quarter keeps your doctor alive.

If we allow ourselves, unwittingly or not, to do something that causes a health issue, then it is only natural for us to expect, sooner or later, to suffer the consequences.

Many medical researchers and nutritionists claim that the primary cause of most human diseases are the foods that are commonly consumed, and that it is the consumption of these foods that leads to the life-shortening diseases of heart disease, cancer, diabetes and the neurological diseases. It is my belief that this claim will be proven to everyone's satisfaction in the years to come, impacting many popular beliefs regarding age-related health issues. An example is the degenerative disease of arthritis, which many nutritionists, including Dr. Ann Wigmore, believe is caused by harmful dietary practices.

"Healing is no accident. All nature heals itself when causes are removed and the conditions of health supplied." - Dr. Herbert M. Shelton.

According to nutritional experts, local treatments are harmful to the entire body, not just the part or parts that may be displaying the symptoms. They contend that almost all attempts made by mainstream medicine to cure diseases are not successful in a true sense, but can, at best, only forestall the diseases from taking over and killing the patients in the short term.

Medications can temporarily relieve pains and inhibit inflammation, but they can weaken the immune system, and are often habit forming. Many of the published medical books and articles available to us about human diseases reveal that patients must often cope with the toxicity of the substances that are prescribed for their illness as well as the illness itself.

As powerful as the medical profession is, it seems to have eyes that cannot see and ears that cannot hear in its quest for cures for diseases. Because of this, we are in danger of being continually affected in adverse ways by medical treatments as time goes by. As emphasized throughout this book, when confronted with a health issue, we should do everything in our power to resolve it without resorting to doctors. When the cause and effect relationship between a health issue and what we are doing to ourselves becomes apparent to us, the best solution will then be known, as well as the proper steps that are needed to put it into effect.

Healing does not come from drugs. It comes from natural bodily processes, such as self-cleansing, or detoxification, that remove accumulated toxins, obstructions and acidosis from the body, from rest, and from providing the body with the nourishment it needs.

The beacon of warning, the clarion call for proper action, has been sent out. We need to become more our own doctors than ever before, and the sooner it starts the better it will be for us, to avoid the unnecessary suffering associated with medical treatments that are not designed to heal.

Furthermore, as evidenced in hospitals throughout the world, many patients of chronic and acute diseases die despite the treatments they receive, or even because of them. In 2014, in the United States and Europe, prescription drugs were the third leading cause of death, after heart disease and cancer. In 2016, a John Hopkins University study indicated that the third leading cause of death in this country was medical errors.[67] [68]

As discussed in Appendix III about the lessons learned from the COVID pandemic, we now have evidence that many of the lives lost to COVID-19 were lost needlessly and avoidably by vaccines and inaccurate medical diagnosis.

These things have led many people to seek second or even third opinions about their health issues. Blind faith in the medical profession has eroded since earlier days, but many are not finding suitable alternatives, nor have they gained sufficient knowledge about foods and nutrition to convince themselves that dietary change is necessary and the need to take charge of their own health is paramount.

Mainstream medicine spends billions of dollars of federal, state and private funds each year trying to find cures for diseases,

---

[67] https://pubmed.ncbi.nlm.nih.gov/25355584/.

[68] https://www.vekluryhop.com/important-safety-information/?gclid=Cj0KCQjwqhttps://www.hopkinsmedicine.org/news/medicine.org/news/media/releases/study_suggests.

especially COVID-19 and the nominal top 10 deadliest diseases in this country. But cures for these diseases remain elusive.

Despite continuing assurances from medical research labs that cures are in the offing (which, by the way, has been going on for quite some time, especially for the top 10 diseases), it is doubtful whether cures will be realized as long as the underlying causes of disease are believed to be foreign substances or organisms.

To effectively deal with any disease, the underlying cause, or causes, of the disease must be known. Mainstream medicine's attempts at curing diseases have not met with success in the vast majority of cases, including COVID-19 and the top t0 leading causes of death in this country, which strongly suggests that the causes that are commonly attributed to diseases are not really the underlying causes at all, but only contributing causes or factors that can localize a disease to certain parts of the body.

It appears that disease will remain, at least for the present, somewhat of a mystery to mainstream medicine, maybe as much of a mystery as disease was to the "medicine man" of yesteryear. Having directed its energies and skills primarily at the suppression of the symptoms of diseases, mainstream medicine appears to have missed the mark.

I am convinced that the key to recovering from a disease and preventing disease from taking root in the human soil of the body in the first place is self-education about foods and nutrition.

Without a sufficient knowledge about foods and nutrition we remain wholly ignorant of how to achieve optimum health. The knowledge needed is best obtained by reading books and utilizing other learning tools, such as the Web, to discover the truth about these vital subjects.

The Nutritionist View

Nutritional experts do not believe that "germs" or bacteria and viruses are the primary causes of disease. Rather, it is the presence of undigested and uneliminated rotting food waste accumulated over time in the tissues and organs of the body due to eating wrong kinds of foods and overeating. In their view, these things mean that elimination of the foul material is the only rational means of curing disease.

"In food lies 99.99% of the causes of all disease and imperfect health of any kind. Consequently, all healing, all therapeutics will continue to fail as long as they refuse to place the most important stress on diet." - Arnold Ehret, *Mucusless Diet Healing System.*

Viruses and bacteria are everywhere and exist in everyone, but before they can proliferate and thrive in the body, they must have suitable soil. Nutritionists believe that it is the waste material in the body that affords the germs the suitable soil for them to proliferate and produce the symptoms of disease.[69]

If we allow ourselves, unwittingly or not, to do something that causes a health issue, and in particular a disease, then it is only natural for us to expect, sooner or later, to suffer the consequences.

Why do natural cures work better than standard medical

---

[69] Louis Pasteur (1822-1895) gave the world the germ theory of disease, but he was never able to prove it. On his death bed he stated, "the germ is nothing, the terrain is everything." In other words, the soil of the body is far more important than any pathogens that attempt to invade it.

treatments?

Many nutritionists, including those referenced in this book, contend that almost all human diseases are the result of internal poisoning, typically organ or blood poisoning, that occurs when the tissues of the body become congested with toxic waste material due to constipation that cause acidity and toxicity of the bloodstream. They believe that disease symptoms are manifested as a result of this poisoning, and that they are signs the body gives when it attempts to eliminate the poisons. Malaise and fatigue are two typical signs of internal poisoning, and so are sores and rashes.

In other words, disease is an attempt to rid the organism of the foul material with which it is contaminated. Also, disease symptoms are the outward manifestation of an automatic internal bodily cleansing process. For example, nausea and fever result from the body's natural attempts to rid itself of morbid, toxic material that has accumulated in the tissues. They are not due to "infectious organisms" that invade the body.

The cure for disease, then, is not to treat the symptoms, including the pains that may result from the disease, but the underlying causes of the symptoms.

Again, nutritionists say that disease is the result of intestinal occlusion, the clogging up of food wastes in the large intestine (colon), and that correcting the situation through proper diet so that the number of defecations per day is about the same as the number of meals eaten per day is the only way the body can rid itself of disease, recover and resume a life of health. If the natural self-healing process is repressed through medication (drugs), and we continue the consumption of foods and drinks that are harmful to the body, then the cure is thwarted.

Furthermore, mainstream medicine places emphasis on treating specific, localized symptoms. Modern drugs are prescribed to correct one symptom or another without effectively strengthening the overall state of health. But the body is an integral organism consisting of many parts, with every part connected with every other part. All parts of the body receive the same blood and lymphatic fluid supply to support its needs. No one part can be diseased and the rest be healthy. Similarly, when one or more parts of the body are healed of ill-health, the other parts of the body benefit as well.

I suffered from various physical ailments for years that caused me a lot of pain and suffering. However, in each case, I tried my best to search out the answers to what caused them, and the knowledge I gained allowed me to determine how to cure them. When the conditions were sufficiently understood, I took what I believed were the proper steps to resolve them.

One may wonder, as I did, why we do not all get diseases, such as polio, diphtheria, typhoid or cholera. The answer from a nutritionist's viewpoint is that there is no suitable soil in our bodies in which these diseases can take root. It appears to be the only reason we do not all get diseases. Microorganisms cannot harm us as long as our organism functions in a normal, healthy way. Microbes and viruses are harmless if the body's defensive mechanisms, including the immune system, are in good working order.[70]

"To the degree that foods are used in their completely natural state, without treatment or processing, are they adapted to support life and maintain immunity to disease." - Arnold Paul De Vries.

---

[70] For a discussion of the immune system and the COVID-19 virus, refer to Chapter 1.

What if a person already has a disease? How then may he or she be cured? According to nutritional experts, if a person already has a disease but deprives the microorganisms their food and renders the human soil unsuitable for their growth, then the disease can be cured. From this perspective, there is hope that disease may not only be prevented, but cured through natural means.

Disease should not be looked on as an affliction brought on by fate or Providence, but something primarily, or perhaps even solely, brought on by an ignorance of the unalterable laws of nature, and most notably the laws governing foods and nutrition.[71]

The body's ability to fight illnesses is determined by the health of the immune system. As stated in the COVID chapter, 70-80% of the cells of the immune system reside in the gut.[72] The gut microbiota aid in food digestion and fight off invading microorganisms. Certain diseases, including cancer, repress the immune system and allow increased attacks by unfriendly microbes and viruses. Nutritionists tell us that the best way to improve the health of the immune system is to eat whole plant foods.

"A whole food plant-based diet deals with so many diseases and conditions that you begin to wonder if there isn't just one basic disease cause – poor nutrition – that manifests through thousands of different symptoms...Poor nutrition causes vastly more diseases than the disease care system currently acknowledges; but good

---

[71] If fasting can cure almost all diseases of which the human race is susceptible, then what does it tell us about how diseases and other health disorders originate? Doesn't it clearly reveal that the principal cause of almost all human maladies lies in the foods that are eaten? Does it not assuredly implicate improper diet as the main cause of our many ailments? Fasting is explored further in a subsequent chapter.

[72] https://pubmed.ncbi.nlm.nih.gov/33803407/.

nutrition, in contrast, is a cure for all those diseases and more." - T. Colin Campbell and Howard Jacobson, *Whole, Rethinking the Science of Nutrition.*

According to Dr. Michael Greger's book, *How Not To Die*, antioxidant supplements, such as vitamin C and beta carotene, do not work. The body needs to get its antioxidants from living plant foods. By eating fruits and vegetables rich in beta carotene you can keep your immune system as well as the rest of your body healthy and happy. The fruits and vegetables that are high in beta carotene include tomatoes, apricots, mangoes, carrots, cabbage, broccoli, cantaloupe, green leaf lettuce, kale, mustard greens, pumpkin, red leaf lettuce, spinach, sweet potatoes, turnip greens and winter squash.

Bear in mind, however, that no cure can be expected to work while eating habits are constantly counteracting the cure.

Many famous nutritionists and nutrition-minded medical doctors assure us that human health primarily depends on the health of the large intestine, or colon, as is evidenced by little or no constipation, and the health of the internal organs and the health of the immune system. It is important to eat as many antioxidant-rich foods as we can to strengthen the immune system, neutralize the poisons within the body and stop or reverse toxic buildup in the body.

Chapter 3 Lifestyle Changes

Lifestyle changes for promoting and sustaining health described herein are accomplished through natural means, which ensures that the entire body, not just one or two parts, is benefited. A healthy diet, for example, can affect a multitude of health issues ranging from nutritional deficiency to constipation, and from feelings of sluggishness, grogginess and general fatigue to diseases of the vital organs of the body.

As stated previously, food is the major culprit in disease, causing almost all disease and imperfect health of any kind (99.99% according to the famous German nutritionist Arnold Ehret). It is because of this fact that all healing, all therapeutics fail as long as they refuse to place the most important stress on diet.

But many people are prevented from making even the simplest of lifestyle changes by what I call "psychological inertia," for we are highly conditioned in this culture practically from birth to do whatever is most convenient to remedy health issues. It is why medication is so often the first choice. For example, according to the Web, at least $2 billion are spent each year in the US alone on antacids, and $10 billion worldwide.[73] That's a lot of indigestion!

As previously mentioned, more than 66 percent of all adults in the US take prescription drugs,[74] and the FDA continues to announce recalls on prescription drugs, including antacids like Zantac. For example, a 2020 web article states:

---

[73] https://www.phlabs.com/stomach-acid-its-a-delicate-balance-be-proactive.

[74] https://hpi.georgetown.edu/rxdrugs/#:~:text=A%20vital%20component.

"A cancer-causing substance known as NDMA has been repeatedly found in one of the most popular antacid drugs in the United States."[75]

Nevertheless, since drugs do not possess the ability to heal since they do not address the root cause or causes of any ailment, no lasting improvement can be expected from them.[76] [77] [78] [79] [80]

A healthy lifestyle is more effective than drugs or medical intervention such as hospitalization and surgery in treating *any* disease, and do not have the undesirable side effects of medical treatments.

Many doctors have not been trained in nonpharmacological disease management, as discussed in the books by leading nutritionists that are referenced in this book. In addition, few people are familiar it. But it is the most effective way to remedy or control a disease as well as practically any other health issue. Some of the nonpharmacological things that improve health can literally transform a person into a vibrant, healthy brand new creation, free of sickness and disease.

The Changing Tide of Disease Treatment

People of every age, gender and ethnic group have benefitted

[75] https://www.wired.com/story/the-fda-announces-two-more-antacid-recalls-due-to-cancer-risk/.

[76] https://www.pritikin.com/your-health/health-benefits/lower-blood-pressure/1220-salt-key-player-in-resistant-hypertension.html.

77 https://www.ncbi.nlm.nih.gov/pmc/articles/PMC2991739/.

[78] https://scopeblog.stanford.edu/2017/10/26/too-high-for-high-blood-pressure-lifestyle-changes-are-the-most-effective-and-safest-drug/.

[79] https://www.medicalwesthospital.org › high-blood-press.

[80] https://www.medicalnewstoday.com › articles › lifestyle.

from simple but important lifestyle changes, the same changes that may have been ignored or neglected previously by the individual in his or her life. A good example is Robert E. Kowalski, who ate the Standard American Diet (SAD) for years before suffering a heart attack at the age of 35, followed by multiple bypass surgeries, and who took blood pressure mediation for years thereafter, but who discovered that most of the medication he was prescribed could be dropped by making some simple but important lifestyle changes. He later wrote a book about it.[81]

"He is the best physician that knows the worthlessness of the most medicines." - Benjamin Franklin.

We are now experiencing a heightened awareness in this country and in the world of the many health hazards that are associated with eating certain foods. The traditional, or standard American, diet is the most commonly practiced diet in the Western world. It consists of high intakes of cooked foods, including meats, dairy products, eggs, refined grain products, jarred, canned and bottled fruits and vegetables, fractionated oils, refined sugar and refined salt. The diet is known to cause many of the diseases and health issues that are prevalent in the world today.

The meat, dairy and grain businesses, which are the chief vested interests in perpetuating the Standard American or traditional American Diet, have influenced federal dietary regulations for decades. It is commonly believed that these interests write the protocols of the USDA. As stated in a dietary guideline article published on the Web, "After all, what is the USDA if not the regulatory body created to ensure that the U.S. agricultural commodities (like corn, soy, and wheat) are profitable?" These

---

[81] Robert E. Kowalski, The Blood Pressure Cure; 8 weeks to Lower Blood Pressure Without Prescription Drugs.

interests are interested first, last and always in profits, not in human welfare.

The vast amount of information now available in the human health field, and in particular in the science of foods and nutrition, can be put into practice to greatly reduce a person's chances of being afflicted by diseases. The importance of adopting a healthy diet, such as one that avoids high cholesterol foods, trans fats and Omega 6 vegetable oils, with their dire effects on human health, cannot be denied, and continues to be as the number one behavioral change for health in general, regardless of the health issue, and particularly for those with high blood pressure.

Next to diet comes exercise. Then weight control. In fact, just about anything proven conducive to health can be included in an effective blood pressure management program, regardless of the particulars of an individual's condition and on time spent on medications.

Perhaps the most important factor in the effectiveness of lifestyle changes for health and longevity is the level of commitment of those who try to incorporate them. If a person is convinced that health is in their best interests to maintain and protect, and make it a top priority in their lives, and are also convinced that true health is best attained by their own efforts, then their level of commitment is likely to be high and they will have the best chances of recouping from a health issue like high blood pressure.

"Always bear in mind that your own resolution to succeed is more important than any other thing." - Abraham Lincoln.

The best way to become convinced that health is best attained by one's own efforts is by reading about foods and nutrition, especially the books written by well-respected nutritionists and nutrition-minded medical doctors, such as Nathan Pritikin, Dr. Michael Greger, Dr. Caldwell Esselstyn, Jr., T. Colin Campbell, Dr. Ann Wigmore, Arnold Ehert and Norman W. Walker.[82]

Remember that if the attainment of genuine health required no knowledge or effort or discernment whatsoever on the part of people, then everyone would be remarkably healthy. But even if you are only half-hearted about lifestyle changes and don't stick with them for very long, you still stand to benefit from the experience of having tried them.

If you're taking medication for any health issue, like high blood pressure, then as your condition improves with lifestyle changes, you may find, as many have, like Robert E. Kowalski, that you can gradually withdraw from it and eventually live without it.
If you are not currently on medication, but know that you suffer from some ailment, then lifestyle changes should be tried before taking medication, since many people have found simple lifestyle changes to have a profound effect on their condition.

The following lifestyle changes for true health are endorsed secondarily (after medication) by many mainstream medical doctors and other health care professionals, but are primarily endorsed by alternative medical practitioners such as those cited and footnoted above, and also by many online sources such as those given here.[83] [84] [85] [86]

---

[82] Dr. Michael Greger, How Not To Die, T. Colin Campbell, The China Study, Dr. Caldwell Esselstyn, Jr., Prevent and Reverse Heart Disease, Dr. Ann Wigmore, Be Your Own Doctor, Arnold Ehret, The Mucusless Diet Healing System, Norman W. Walker, Become Younger.

[83] https://www.mayoclinic.org/diseases-conditions/high-blood-pressure/in-depth/high-blood-pressure/art-20046974.

[84] https://www.heart.org/en/health-topics/high-blood-pressure/changes-you-can-make-to-manage-high-blood-pressure.

1. Dietary change. A change in diet is the most recommended lifestyle change for reversing and healing a medical condition. Why? Because of what has already been explained about the causes of disease, that they are principally (99.99%) diet related. It is by far the best solution that anyone can incorporate for safeguarding their health and healing a disease or other health issue.

People in this country should be the healthiest people on earth. It certainly seems that way since there are now more varieties of health-promoting and sustaining foods available to us than ever before, and many of them are organically grown.[87] But we are not the healthiest people because of our preferences for artificial, man-made foods and drinks, our giving way to cultural norms and traditions about foods, and our unwillingness, and even refusal, to abide by Nature's laws that govern health and well-being.

"We must close our ears and our minds to the false prophets posing as "experts," who ignorantly recommend "man-made foods" that are slowly but surely hastening our end." - Professor Arnold Ehret, *Thus Speaketh the Stomach and the Tragedy of Nutrition.*

Natural laws govern our world. They include the laws of physics that encompass the mysterious forces of electromagnetism, gravity and the nuclear forces, and the laws of biological processes, including life itself. Natural laws are always in effect, and they apply to everyone in every generation.

---

[85] https://www.healthline.com/health/high-blood-pressure-hypertension/lower-it-fast.

[86] https://www.medicalnewstoday.com/articles/318716.

[87] Organically grown produce does not have the highly toxic herbicides and pesticides that are sprayed on conventionally-grown crops, and it is not grown in mineral-starved soils which are used to produce conventional crops.

All creatures seem to be perfectly attuned and adjusted to these laws, except for man. We are continually challenging or testing Nature in one way or another. Our free will, which other creatures do not have, and which sets us apart from the other creatures perhaps more than anything else, enables us to question the validity of natural laws. But Nature has a way of punishing and even eliminating people who break her laws.

"Among the many thousands of species of creatures living on the earth, only humans and some of their domesticated animals (dogs, cats) try to live without food enzymes. And only these transgressors of nature's laws are penalized with defective health." - Dr. Edward Howell, *Enzyme Nutrition*.

Whole plant foods are different in many ways from other foods. Apples, when planted in the soil, produce additional apple trees. Raw nuts planted in the ground produce other nut trees. Even a harvested potato when planted yields at least another potato plant. But many of the foods that are consumed on ordinary diets have been devitalized by heat treatment. Cooking destroys the life force properties that are in whole plant foods. Plant a cooked bean or tomato or a roasted nut in the ground and it will not grow. Cooked foods, including refined and processed foods, do not promote or sustain health, but are harmful to the body.

The law of cause and effect is one of Nature's laws. It is observed in many clinical trials and studies that are conducted on human health and nutrition each year that show a direct relationship between the diseases of humankind and diet. Some of these trials and studies are documented in articles of the scientific and medical professions, many of which are available on the Internet, and some are cited in this book as well as in the books that are listed in the Bibliography.

If you don't want the effect, do something about the causes.

"The present ignorance of the laws underlying normal health is now, in this century, the greatest of all the past centuries, and is evidenced by the deterioration of the so-called civilized people health-wise." - Arnold Ehret, *Physical Fitness Through a Superior Diet, Fasting, and Dietetics.*

2. Daily exercise. Exercising moderately for 30 minutes a day is extremely important for health, and can lower a high blood pressure condition by 5 to 8 mm Hg. Long walks, cycling, swimming and lifting weights helps not just one or two parts of the body, but all of its parts. One needn't worry about exercising temporarily raising blood pressure during exercise, which it does, because it lowers it on an average basis. Also, the time it takes to recover from increased heart rate experienced during exercise decreases with as one becomes more fit.

"People who are physically fit recover quickly from exercise. Their pulse rates and heart rates return to baseline much faster that those who are less fit." - Carol Tavris, Anger, *the Misunderstood Emotion.*

3. Lose weight. It is well known that weight gain affects a person's susceptibly to getting disease. It is virtually impossible to gain much weight on a whole plant food diet, and it is unheard of on the raw vegan diet which consists entirely of raw plant foods.

4. Reduce stress. Chronic stress raises blood pressure. Medication is widely prescribed and taken for nervous conditions, but no drug can cure a nervous disorder. You can cope with stress through medication, but you cannot get rid of it that way.

The best way to handle stress is through rest and relaxation, doing the things that Nature compels us to do. The remedy is so simple that it is generally overlooked, but it's true.

Chronic stressors include repeatedly spending time with someone you dislike, working for someone you dislike, and experiencing road rage. Avoiding such things, or getting away from other things that cause stress levels to spike lowers blood pressure and promotes overall health and longevity.

Another form of stress is noise. Noise is unwanted sound, and its effect on people differs from person to person, but any noise that is upsetting to where it arouses anger will raise blood pressure, and if prolonged will cause chronic stress. Like other stressors, noises are best avoided wherever possible.

Prayer and meditation are known to relieve stress, bitterness, anger, disappointment, heart break, old grudges and frustration, all of which contribute to bowel disfunctions like constipation and other maladies such as high blood pressure. Both practices have a cathartic effect on the mind.

Independent clinical studies have shown that meditation lowers blood pressure significantly. Researchers have found that regularly practicing techniques like Transcendental Meditation helps in numerous ways. Consistent practice can lower systolic pressure by more than 10 mm Hg and diastolic pressure by more than 6 mm Hg.[88]

If it were not for taking these "time outs" from daily struggles, we would be swamped and continually irked by the many stresses that are imposed on us. But when we practice prayer or

---

[88] Robert Rowan, M.D, Control High Blood Pressure Without Drugs.

meditation, a profound peace descends upon us. It leads not only to the momentary relief of worry and stress, but to improved mental health and emotional stability. It relieves us of the oppressive urgency of the present, the regret of the past and the fear of the future.

5. Stop smoking. If you smoke, stop. Cold turkey is perhaps the best way. It's how I stopped after smoking 2-3 packs per day for over 30 years when I became convinced that it was doing me irreparable harm. Smoking reduces the oxygen available to the lungs, replacing it by carbon monoxide, carbon dioxide, and cancer-causing chemicals that are sprayed on the tobacco leaves during their growth, and, if paper cigarettes are used, cancer-causing chemicals are sprayed on the cigarette paper, and all these things cause lung cancer and heart attacks.

Effort is required to accomplish anything worthwhile in life. The dictionary is the only place where success comes before work. Surely, nothing worthwhile is ever accomplished without effort, and the things that require the most effort are often the most rewarding and worthwhile of all.

It takes work to get things right. You can never escape that fact. Practically every aliment of which human beings are prone can be effectively prevented or controlled through lifestyle changes.

For some people, it takes a combination of medication and lifestyle changes to remedy a health condition, but no matter what route you end up taking, remember that if optimum health and longevity required no knowledge or effort or discernment whatsoever on the part of the individual, then they would be easy to come by and everyone would be wonderfully healthy. But all things are given to us at the expenditure of effort.

Chapter 4 More About Diet

"Let food be thy medicine and medicine be thy food." – Hippo-
crates.

Previous chapters have pointed out the importance of adopting a
whole plant food diet for health. This chapter discusses important
specifics about a whole plant food diet and elaborates further on
its importance for health.

But everyone is different, right? Everyone requires different foods,
don't they? Well, it is true that not everyone has the same
proclivity, or liking, for foods, or the same nutritional
requirements. But the body requires similar things for health and
well-being, and we are all very much alike in numerous ways,
having the same basic needs, the same general physiology, the
same susceptibility to biological disharmony, the same
susceptibility to many illnesses, and we share many of the same
stresses, worries and fears that can influence our state of health.
Nutritionists, whose voices have not been heard as much as those
of ordinary diet proponents, have been telling us for years that ill-
health and diseases of all types spring most often from one
common source, namely, poor dietary practices.

The best diet – for anyone – is the one that meets the energy and
nutrient needs of the body and produces the least negative
effects, such as toxicity, acidity and constipation. For most people,
it means a diet consisting mostly of fresh, whole fruits and
vegetables to meet the body's nutrient needs, and enough
concentrated foods, such as nuts and seeds, to meet the body's
energy needs. However, just as exercise must be tailored to the
person, so it is with whole plant foods since not every kind of plant
food is best for all people.

Nutritional scholars like Dr. Joel Fuhrman, author of *Eat to Live*, tell us that there is more evidence that meat and dairy products cause disease than smoking causes lung cancer.

Refined and processed foods are responsible for a large number of health disorders, ranging from headaches to the nominal top 10 leading causes of diseases in this country, and most diets permit these foods to be eaten.

Almost every kind of food that comes in a sealed package (bag, box, can, jar, bottle, etc.) has been refined and processed, meaning, among other things, that it has been heat-treated. The enzymes in the foods have been destroyed by heat, and superfluous and harmful substances have been added as discussed previously in this book.

Nothing that man can do to foods, such as high temperature heat treatment, the addition of preservatives, artificial colors and sweeteners, antibiotics (in animal products), and other refined and/or man-made ingredients, can improve the food value of foods over their living food counterparts. To think that it can is ludicrous, since anything done by man to foods (other than necessary harvesting or shipping) alters or adulterates the foods in some way.

"The adulterated, unnatural, false, man-made foods of present-day civilization are the underlying, physiological causes of all evils to which humans are prey – especially of all kinds of diseases. Health will not return, nor can it be regained, through drug remedies or other various treatments offered by medical science, since supreme, absolute, paradisiacal health is ruled by the laws of diet" - Professor Arnold Ehret, *Thus Speaketh the Stomach and the Tragedy of Nutrition*.

Dr. N.W. Walker spent most of his life exploring man's capability to extend life. His books are referred to several times in this book since they provide much insight into what foods should be eaten for longevity and optimum health. According to Dr. Walker, the foods that people should eat for longevity are whole plant foods. Whole, or raw, plant foods are the most conducive of any foods to longevity because they contain an abundance of life force properties and enzymes and do not contain man-made and man-altered ingredients. He lived to be 99 according to the Web, but some sources say 109.

A healthy diet is one that is low in saturated fats and Omega-6 oils, and high in natural fiber, vitamins, minerals and other nutrients that found easily assimilable in whole plant foods as opposed to supplements or man-made foods. According to the government's Dietary Guidelines for Americans, people should get most of their nutrients from food and beverages rather than supplements.[89]

The body is much wiser than we may think it is. It affects the mind so significantly that it may be said to control the mind. Eating a whole plant food diet results in genuine health, which profoundly affects the mind by increasing mental stamina and clarity.

One of the major obstacles in the way to successfully attaining a whole plant food diet is what I call the curse of "food convenience." Our society is built on fast foods and fast drinks. We need to break away from these things. It requires shifting our focus to healthy foods and drinks, those with health-appeal rather than taste-appeal. It means taking the time required to shop at local food stores for our foods, even though doing so typically takes only minutes more than it does to drive through a fast food

---

[89] https://ods.od.nih.gov/factsheets/Magnesium-Consumer/.

place. This may be a major shift in thinking for many people. It's hard to break old habits. However, they were formed by repetition and they can be broken by repeatedly refraining from them.

"We must close our ears and our minds to the false prophets posing as "experts," who ignorantly recommend "man-made foods" that are slowly but surely hastening our end." - Professor Arnold Ehret, *Thus Speaketh the Stomach and the Tragedy of Nutrition.*

Numerous clinical studies have shown that ordinary diets, such as the Standard American Diet, creates unhealthy food cravings and chaotic, haphazard food combining[90] which, in turn, produce sickness. But a diet of life-giving foods, together with fresh air, clean water and exercise, enable the body to heal itself of practically any health issue, including high blood pressure, cancer, heart disease, stroke and the many lesser ailments of which humankind is susceptible.

The cooking of foods at home and in restaurants, and in the food factories that manufacture canned, bottled and jarred foods, reduces the food value of the foods by altering their chemical properties and destroying important food components, such as enzymes and nutrients, including vitamins and antioxidants.

If food is browned by heat treatment, such as by broiling, baking or deep frying, dangerous chemicals are created, such as advanced glycation end-products (AGEs), which have been known to cause diabetes and heart disease. The body will not tolerate continuous

---

[90] The Web adequately covers many aspects of proper food combining. A good website on proper food combining at the time of this writing is: https://www.acidalkalinediet.net/correct-food-combining-principles.php.

abuse without exacting a penalty. More about the dangers of eating cooked food is found in the book *A Christian Diet*.[91]

Many nutritionists and nutrition-minded medical doctors, including those referenced in this book, believe that the neglect of the foods that God intended for us to eat has brought about the woeful health conditions that are so prevalent in the world today, including almost all of the diseases known to afflict man. Anyone can live healthier by eating natural foods. If used properly, they work in wonderful ways to cure health issues, including diseases, restore long-lost energy and prevent disease from taking root in the body, and each of these things work to increase longevity.

Optimum health requires an understanding of foods and nutrition that many people do not have. Health is not for the asking. If it were, then everyone would be extraordinarily healthy. Rather, health must be gained, it must be earned. In order for the body to rid itself of diseases and other health issues, health-damaging food habits must be changed.

However, generally speaking, people are unwittingly ignorant of what foods make them well. While there are scores of new dietary fads coming out each year that tempt many to change their diets, there is, in my experience, only one sure way that ensures long-lasting benefits, and that is the whole plant food diet. Adopting a whole plant food diet puts you on the road to optimum health.

The Plant-Based Diets

For comparison purposes, the plant-based diets are as follows:

---

[91] The book is listed in the Bibliography.

Vegetarian Diet – The vegetarian diet is a diet that consists mostly of plant foods. It minimizes meat and dairy products but can include some fish, and/or some dairy and poultry such as cheese and eggs. The emphasis of the vegetarian diet is on eating mainly plant foods but also cutting back on animal-based foods. Both cooked plant and animal-based foods are allowed on the vegetarian diet.

The Dietary Approaches to Stop Hypertension (DASH) eating regimen, which emphasizes eating fruits and vegetables, is a vegetarian diet that's been popular with people with high blood pressure. It provides three times more potassium than the average American diet. It has been shown to lower systolic blood pressure by an average of 5.5 mm Hg and diastolic blood pressure by 3.0 mm Hg.[92] [93] Researchers have found that a diet like the DASH diet not only reduces blood pressure, but as one would expect, it is also effective in preventing cardiovascular disease and kidney disease.

"The great thing about this finding is that we aren't talking about a fad diet. This is something that many physicians already recommend to help prevent chronic disease."[94]

Vegan Diet – The vegan diet is somewhat like the vegetarian diet but it goes further to exclude all animal-based foods, even fish, cheese and eggs, although modified versions of the vegan diet may allow them. The vegan diet allows cooked plant foods to be eaten.

---

[92] The DASH diet was developed in the 1990s to help remedy high blood pressure conditions.

[93] https://ods.od.nih.gov/factsheets/Potassium-HealthProfessional/.

[94] https://www.jhsph.edu/news/news-releases/2016/diet%20-designed-to- lower-blood-pressure-also-reduces-risk%20of-kidney-disease.html.

Raw Vegan Diet – The raw vegan diet is somewhat like the vegan diet in that it excludes all animal-based foods.[95] But it also excludes all cooked foods, which includes cooked plant foods.

A vegan or even a vegetarian diet can significantly reduce arthritic inflammation and pain. Even Web articles state this. However, a raw vegan diet, which goes further than both of these diets in excluding harmful foods, does it even better. Modern nutritionists, including those listed in the Bibliography, agree that the raw vegan diet, together with drinking distilled water and fasting, dissolves accumulated inorganic minerals and toxins that are deposited in the joints and tissues of the body better than any other health regimen.

There is no need to discuss diets other than plant-based diets, including the many fad diets that come out each year, for example, the various high-protein, high-fat and low-carb diets, since animal products and cooked foods, which are part of these diets, are harmful to the body, as maintained by many nutritionists and as explained in this book.

Raw plant foods are foods as found in Nature. They enable the body to heal itself of diseases and attain optimum health. The reason they do this is because of the natural life force that is in

---

[95] Bee products are often contested as not being raw vegan food since they come from bees, which are animals. However, there are many nutritionists, among them Ann Wigmore, Norman W. Walker, Herbert M. Shelton and David Wolf, who advocate including bee products in the raw vegan diet. In my opinion, this issue is, in the overall picture of things, a minor point of contestability, especially when the health benefits of bee products are considered. In any case, each of us must decide how we stand on this issue. Like any other food types, if bee products do not work for you, then they should be avoided.

these foods. This life force comes from the sun and is converted by plants into energy that humans and animals can utilize. When we eat raw (uncooked) fruits and vegetables, including green leafy vegetables, we get the precious life force that God and Nature intended for us to receive.

Raw plant foods by their very nature promote optimum health and longevity. They give you the zest and energy you always wanted but never could quite obtain. They make you feel good about each and every day of your life. The raw vegan diet has been shown to literally transform a person into a new person. Other diets cannot do this. The more raw (whole) plant foods you eat, the more antioxidants, enzymes and other vital nutrients you will get.

Enzymes have life force properties that Nature intended for us to receive. They support all bodily functions and contain the vitamins and nutrients the body needs for optimum health. They are discussed in detail in the book, *A Christian Diet*.

If you are not currently on a whole plant food diet, I recommend transitioning to the raw vegan diet by first becoming a vegetarian, then a vegan and then a raw vegan. The reasons for this are explained in the book, *A Christian Diet*.[96]

You begin to take more time preparing meals, the meals that are right for you, and this continues as you age. The curse of food convenience will no longer be on you.

Each kind of living plant food has its own unique blend of nutrients and life-giving properties. We can benefit from eating many types of raw plant foods. I believe the vast variety of plant

---

[96] The book is listed in the Bibliography.

foods in the world have been given to us for a purpose, and that we can discover that purpose by becoming familiar with these foods.

An aspect of a whole plant food diet that strikes me as being perhaps the most interesting is that it doesn't leave you guessing about whether your health is going to improve. Your health does not stall or level-off and then decline on the diet, as it does on other diets. It continues to improve.

What are the Best Foods for Health?

Some of the foods that are highly recommended by nutritionists and nutrition-minded medical doctors are discussed in this section. The foods they recommend avoiding are those found in ordinary diets, such as meats and dairy products.

Fruits and Vegetables

Raw fruits and vegetables should be eaten regularly. They digest efficiently because their enzymes have not been destroyed by heat.

Fruits and vegetables are best eaten when they are ripe. If eaten in their typical store-bought, un-ripened condition, stomachache or some other discomfort is likely. Also, extra energy is required to digest them, and this energy is taken from the energy reserves of the body when it could be used for other purposes, such as healing and self-cleansing.

Fruits and vegetables sold at local food stores are typically shipped-in from distant locations, such as foreign countries, and purposely arrive in an un-ripened condition in order to retard spoilage and prolong shelf-life. To ripen store-bought fruits and vegetables, just set them on the counter tops at home until they are ripe; it usually takes several days, depending on the fruit or vegetable. For example, bananas are typically sold green or partly green in color, unless you happen by the fruit stand right before the bananas are replaced. Ripe bananas are speckled or streaked brown in color, which typically takes several days.

Cucumbers are ripe when they are easily flexed. Avocados are ripe when they yield to gentle pressure. Green chilies and jalapenos are ripe when they turn orange or red, do not eat them when they are green. Lemons, limes, oranges and pears are ripe when they are aromatic. Same for red and yellow peppers.

Some exceptions to this are apples (all varieties) and root vegetables (for example, carrots, beets, turnips, radishes, potatoes, onions, garlic, etc.). Apples and root vegetables do not ripen to any significant extent after they are picked.

Fruits are nutritious and stimulating, and have many healing qualities. They act to cleanse and energize the cells of the body. I consider fruits to be the ideal food for humankind, and we are blessed indeed to have such a plenteous supply.

"Fruits alone, even of but one kind, not only heal but nourish perfectly the human body, eliminating entirely the possibility of disease." - Arnold Ehret, *Mucusless Diet Healing System*.

Fatty fruits include avocados and olives. They contain healthy unsaturated fat. Some nutritional experts consider avocado a Superfood, although it is not classified as such. The avocado contains a substantial amount of monounsaturated fats, phytosterols and antioxidants like vitamin E, vitamin C, and carotenoids. It is also high in beta-sitosterol (95 mg per medium-

sized avocado) which is known to assist in relieving prostate disorders, such as benign prostatic hyperplasia (BPH).
Plain avocado can be used like butter on raw vegetables such as cabbage, broccoli, cucumber, cauliflower, carrots, asparagus, green onions, tomatoes, chili peppers and celery – with great results. But it's mainly fat, so use avocado in moderation.

The vegetables that have the most tightly compacted layers are some of the most nutritious. They include red and green cabbage, leeks, broccoli, bok choy, green onions, lettuce and celery. Leafy green vegetables, or leafy greens, include spinach, arugula, chard, kale (several varieties), mustard greens, collard greens, turnip greens, parsley, cilantro, lettuce (several varieties), celery greens and dandelion.

"Celery can lower your blood pressure. It actually contains God-designed phytochemicals like apigenin and 3-N butyl phthalide which can relax blood vessels, providing the benefits of calcium channel blockers without the proven brain shrinking as well as the diuretic effects and hormonal effects like lowering noreoinephrine." - Dr. Sherry A. Rodgers, *The High Blood Pressure Hoax*.

Leafy greens are commonly considered vegetables, but I agree with Victoria Boutenko in her book, *Green for Life*, that greens should be their own food group. Greens are the only foods that combine well with all other foods, including other greens. You can't say that about many fruits and vegetables or nuts and seeds.

The chlorophyll in greens strengthens the immune system, helps to detoxify the body and improves digestion. Chlorophyll is rich in antioxidants, minerals, vitamins and readily assimilated enzymes. The chlorophyll in green plants is what converts sunlight into chemical energy, and this energy is made available to us when we eat greens.

Eating a variety of greens ensures that we receive all the amino acids we need in our diets. In his book, *The China Study*, T. Colin Campbell states that plant protein is the healthiest type of protein because it allows for slow but steady synthesis of the proteins.

"Greens are the primary food group that matches human nutritional needs most completely...Chlorophyll is liquefied sun energy. Consuming as much chlorophyll as possible is like bathing our inner organs in sunshine." - Victoria Boutenko, *Green for Life.*

Carrots are high in beta-carotene which is converted to vitamin A in the body. The word "carotene" is derived from the Latin word for carrot, "carota." Nutritional expert Dr. N.W. Walker in his books (see Bibliography), states that raw carrots have all the elements and vitamins that are required by the human body. It could just be that the "lowly carrot" is capable of making up for many of the nutritional deficiencies in the world today.

Carrots are non-starchy vegetables.[97]

Red beets are good for the blood. They lower blood pressure. They improve athletic performance. They are one of the highest of oxygenating foods. Marathon runners are partial to them because they increase their endurance. After consuming red beets, it takes less energy to run a race. This makes red beets important for elderly health, since studies have shown that there is a decline in maximal oxygen consumption with age. In addition, red beets have a high acid-binding rating.

---

[97] S. H. Shepherd, A Christian Diet.

The Creator has blessed and enriched the earth with a wide variety of life-promoting whole plant foods. We have spent too much time and effort learning about the foods of the animal kingdom and how to prepare them, and look what it has cost us in terms of health and lifespan. It is time we learned about the foods of the plant kingdom and the many tasty and nutritious meals that can be made from them.

Whole plant foods are different in many ways from other foods. Apples, when planted in the soil, produce additional apple trees. Raw nuts planted in the ground produce other nut trees. Even a harvested potato when planted yields at least another potato plant. But many of the foods that are commonly consumed have been devitalized by heat treatment. Cooking destroys the life force properties in foods. Plant a cooked bean or tomato, or a roasted nut in the ground and it will not sprout. Cooked foods, including refined and processed foods, do not promote or sustain health, but are harmful to the body.

We live in a green world. Green is the color of nature, the symbol of youth and growth. Green plants, from lowly grass to lofty trees, together with water, hold the key to life on earth. Even the oceans contain many green plants.

The plant kingdom is a vast reservoir of energy. Plant foods capture the sun's rays and convert their energies into foods for man and beast. Animals cannot convert the energies that are in sun rays directly into energy that can be utilized by the body. The chlorophyll in the leaves of green plants converts sunlight into chemical energy, and we receive this chemical energy when we eat raw plant foods. All animals, including humans, depend on plants to do this for them.

The different wavelengths, or energies, of the sun's electromagnetic energy spectrum are believed to be one of the reasons why there are so many different varieties of plants on the earth. It is believed that some plants are more receptive or attuned than others to the different wavelengths. Nutritionists tell us that the more varieties of plant foods we eat, the more we benefit from the different energies.

Each kind of living plant food has its own unique blend of nutrients and life-promoting properties. We can benefit from eating many types of raw plant foods. I believe that the vast variety of plant foods in the world have been given to us for a purpose, and that we can discover that purpose by becoming familiar with these foods.

The foods that are best for us are foods that have their life-giving properties intact, that enable the body to withstand the onslaught of diseases, that provide the energy needed for an active and productive life and that promote and sustain optimum health. Nutritional experts agree that a whole plant food diet, such as the raw vegan diet, minimizes bodily toxicity and acidity and rids the body of its toxins better than any other diet.

"Uncooked foods will supply not only all the necessary vitamins and minerals, but also all the enzymes and easily digestible natural starches and proteins needed for healthy functioning of the body." - Paavo O. Airola, N.D., *There is a Cure for Arthritis*.

Man is Best Suited for Eating Whole Plant Foods

In *Fruits and Farinacea – The Proper Food of Man*, John Smith tells us that based on all accessible sources, our progenitors were frugivorous, i.e., fruit eaters. Both anthropological studies and studies of how the human body functions support this conclusion. "Meat is not man's natural food, since he is not either a

carnivorous or an omnivorous animal. Every argument drawn from comparative anatomy, from physiology, from chemistry, from experience, from observation, and, when rightly used, from common sense, all agree that man is not a meat-eating animal. He can never be as healthy under the prevailing "mixed" diet as he would if he were to follow the dictates of Nature and live on his natural food – fruits and nuts, eaten in their uncooked, primitive form. Every element the system needs can be shown to be present in these foods, in their proper proportion, while, being live foods instead of mere "dead ashes", which is all the cooking process leaves, they will be found to supply a degree of vital life and energy which no cooked foods ever supplied or could supply."
- Hereward Carrington, *Vitality, Fasting and Nutrition*.

Human beings, in many key physiological ways, are not like other animals. We have hands with opposable thumbs, non-claw-like nails, teeth that are not suitable for tearing hide or flesh, or breaking bones, but rather for grinding plant foods, and long, not short, digestive tracts including 20-30 foot long intestines that are ideally suited for digesting fiber-rich foods like fruits, vegetables, nuts, seeds and grains.

David Wolfe in his book, The *Sunfood Diet Success System*, includes as Appendix A, an "Anatomy Chart" that identifies 17 physiological ways in which human beings are ideally suited for eating a plant-based diet. Websites also support this conclusion, as can be seen by searching on "Are humans frugivores?"

These studies show that humans are naturally suited for picking, chewing and digesting plant-based foods. Chimpanzees, which are very similar to humans physiologically, subsist almost entirely on fruits and greens.

According to the Bible, the first people on earth lived to very great ages. Adam lived to 930 years Methuselah lived 969 years. Prior to the Flood, the average human lifespan was about 900 years. However, immediately after the Flood, when animal food was permitted to be eaten, the average lifespan fell to about 400 years. Later, when Jacob, the father of the twelve tribes of Israel, lived, the average lifespan was only about 150 years.

Based on the latest worldwide statistics from WHO, the average human lifespan is 72 years.

Scientists tell us that the human brain has billions of cells but no more than a tenth of them are ever used. It would seem to make sense if the brain, as well as the rest of the human organism, was designed to last for hundreds of years instead of less than a hundred.

Healthy Eating Habits

Healthy eating habits are formed when we accustom the mind to new tastes and food selections found in living plant foods based on reason and knowledge rather than allowing cultural norms and traditions to tell us what foods we should eat or the emotional pulls that certain foods have on us.

Food cravings are encouraged by the many advertisements in the media for foods and drinks. Despite their oftentimes silky persuasiveness, the products being sold are, for the most part, refined and processed foods and drinks that have been heat treated, chemically altered, devitalized and robbed of precious nutrients. Not surprisingly, many of the ads support the meat, dairy and grain interests, the chief vested interests in our food industry.

Learning the truth versus hype about foods and nutrition is the best way to reach health goals. The knowledge gained during the learning process is priceless, for very few things are as important as good health.

Anyone with a little persistence and willpower can change their eating habits, or for that matter any other habit in life. Voluntary change begins in the mind. Healthy eating habits are formed when we become convinced that dietary change is in our best interests.

However, many people believe that they are, and have been, living healthy lives, and in particular, they believe that they are eating healthy foods, even when they have health issues. But this is belied when the causes of disease and lesser health disorders are examined, as we have already seen.

I grew up at the time when fast-food franchises were just starting out in this country. I was hooked on, and for years ate, the foods of the Standard American Diet. But as I learned more and more about foods and nutrition, and saw the ill effects that the diet had on myself and others, I began eating healthier foods.

By choosing to ignore information that has been gained in the human health field which proves that diseases and a host of lesser human ailments are caused by commonly eaten foods, many of us are sacrificing our health for eating habits and food cravings.

"Once I had an overwhelming Diet Coke craving and went to the 7-11 and got a Big Gulp. The baby inside me shook for hours afterward. (I did vow never to do that again.) My blood pressure, always about 93/55, rose to 120/80. I gained 65 pounds. My ankles swelled up with edema to the point where, as a trick at parties, I'd push my finger in and watch how the depression in my ankles stayed there for 30 seconds. I developed hemorrhoids and terrible blood sugar problems. I once saw a former boyfriend as I was walking through the mall and said hello, but my face was so fat he just gave me a strange look and kept walking. He had no idea who I was." - Robyn Openshaw, *The Green Smoothies Diet*.

But the remarkable truth is that eating habits can be changed, even when food cravings exist, and that everyone is capable of changing their diet in enlightened self-interest.

"I now know how to recognize a healing crisis. I know which foods are truly nutritious and which aren't. This is information that the vast majority of American do not have. It's not just that they lack self-discipline or make poor choices – they truly don't know because of the false and downright injurious education they have received. I know how to heal and build my own body and my children's with nutrition. I know how to cleanse the organs of elimination whenever necessary. Most importantly, I know how to avoid the need for all that, massively reducing our risk of not only degenerative diseases like cancer and heart disease, but also of simple colds and flu, by just eating simply, low on the food chain, every day." - Robyn Openshaw, *The Green Smoothies Diet*.

An awareness of what foods to eat and which foods to avoid, and living on a well-balanced diet that meets the body's nutritional needs makes life a pleasant experience when all around you people are suffering from health issues of numerous kinds because of the foods that they eat.

Fasting

Naturopathic healers and fasting professionals, including those who work in the renowned health institutes and clinics of Sweden, Germany and the US, such as the Buchinger-Wilhelmi fasting clinic in Germany, are keenly aware of how fasting can cure diseases. They witness daily the amazing results brought about by fasting cures, and consider fasting to be the most important curative measure in disease treatment.

Many health advocates and raw foodists regularly fast for days at a time, including Professor Spira (his book is listed in the Bibliography).

Clinical Evidence on Fasting Cures

A vast amount of clinical evidence has been obtained on fasting cures. It has resulted in thousands of publications attesting to the amazing power of fasting in enabling the body to heal itself of diseases and other health issues. They include case histories of "miraculous" cures, some of which are without precedent.

The results of a small sampling of the clinical evidence on fasting cures are given below. For more clinical studies, see the books on fasting listed in the Bibliography.

715 cases of disease were treated by fasting in Dr. James McEachen's sanatorium. The diseases included heart disease, cancer, high blood pressure, kidney disease, ulcers, colitis, arthritis and multiple sclerosis. Remedied or greatly improved by fasting were 29 of 33 cases of heart disease, 20 of 23 cases of ulcers, 3 of 4 cases of multiple sclerosis, 36 of 41 cases of kidney disease, 77 of 88 cases of colitis, 39 of 47 cases of arthritis and all of the cancer and high blood pressure cases.[98]

"Fasting works by self-digestion. During a fast your body intuitively will decompose and burn only the substances and

---

[98] Arnold Paul De Vries, Therapeutic Fasting.

tissues that are damaged, diseased or unneeded, such as abscesses, tumors, excess fat deposits, excess water and congestive wastes. Even a short fast (1 to 3 days) will accelerate elimination from your liver, kidneys, lungs, bloodstream and skin. Sometimes you will experience dramatic changes (cleansing and healing crisis) as accumulated wastes are expelled. With your first fasts you may temporarily have cleansing headaches, fatigue, body odor, breath coated tongue, mouth sores and even diarrhea as your body is cleaning house. Please be patient with your body!" - Paul C. and Patricia Bragg, Water, The Shocking Truth That Can Save Your Life.

Fasting signifies, and in the practice of fasting is found, almost all of the health principles that are needed for optimum health. If fasting can cure almost all diseases of which the human race is susceptible, and it can, then what does that tell us about how diseases and other health disorders originate? Doesn't it clearly reveal that the principal cause of health disorders lies in the foods that are eaten? Does it not assuredly implicate improper diet as the main cause of many of our ailments?

The Importance of Organic Foods

Federal regulations in the US stipulate that organic foods cannot be grown with synthetic (man-made) fertilizers, synthetic pesticides or sewage sludge. As such, organic foods do not contain harmful or potentially toxic substances. These regulations were a long time in coming.

Organic foods are not irradiated,[99] and they are not GMO-modified (i.e., Genetically Modified Organisms). Organic produce

---

[99] Dr. Edward Howell states in his book, Enzyme Nutrition, that the use of radiation to preserve foods results in the wholesale destruction of all the enzymes and vital properties contained in the foods.

will typically have the "USDA Organic" label, but may have the "CCOF (California Certified Organic Farmers) Organic" label which is somewhat less restrictive than the "USDA Organic" label. Other organizations are also USDA certified, such as Oregon Tilth, which uses the "Oregon Tilth Certified Organic" (OTCO) label.

There are now more certified organic growers/farmers in this country than ever before. Accordingly, prices have gone down for organic foods.

We can buy non-organic bananas, lemons, limes, pineapples, avocados, onions and garlic because they require less pesticides in their cultivation, so they should have less pesticide residues on them. Also, we don't usually eat the skins of these foods.

Health is not freely bestowed on anyone, but is given to those who are willing to seek it. There is no magic pill or silver bullet for health. To achieve it, it must be earned.

Chapter 5 Blood Health

"The life of the flesh is in the blood." - Lev. 17:11.

This chapter describes the effects that foods have on blood health and shows how to interpret your own blood condition.

Blood is life. As the poets claim, it is the song of the lark, the blush on the cheek, the spring of the lamb. It is the sacred wine in the silver chalice. Down through the ages, blood has been the price men paid for freedom, and so it is today. Blood is our most preciously guarded possession.

Nutritionists tell us that the quality of the blood starts changing within a few hours after eating a meal. Blood cells, like other cells of the body, are continually being replaced. Old cells are being replaced by new cells. The new cells are constructed from the raw materials that foods and drinks provide. The quality of the blood depends a great deal on the quality of the foods that are eaten. When we eat cooked foods, including refined and processed foods, blood cells are constructed of inferior-quality building materials, materials that are devoid of the life force properties that raw plant foods possess.

"Dead atoms and dead molecules cannot rejuvenate or re-generate the cells of the body. Such food results in cell starvation and this in turn causes sickness and disease." - Norman W. Walker, *Water Can Undermine Your Health*.

Acidity is a blood condition mainly brought on by eating too many acid-forming foods. When we eat foods typical of a cooked meat and/or pasteurized milk diet, which are acidic foods, the blood becomes thick and heavy which causes clogging in the tissues and is known to adversely affect the arteries and lymphatic system, and cause poor circulation and elevated blood pressure. When we eat whole plant foods such as raw fruits, vegetables and leafy greens, the blood's condition becomes normal, which is alkaline, and is not thickened which results in improved circulation and reduced blood pressure.

Several independent researchers have shown that high dietary acid may be linked to kidney disease.[100]

"Animal foods cannot build good blood; in fact, do not build human blood at all, because of the biological fact that man is by nature a fruit eater. Look at the juice of a ripe blackberry, black cherry or black grapes. Doesn't it almost resemble your blood? Can any reasonable man prove that half-decayed "muscle tissue" builds better blood?" - Arnold Ehret, *Mucusless Diet Healing System*.

Almost all raw plant foods are alkaline, or become alkaline in the body. Fruits of the Citrus genus (oranges, grapefruit, etc.) are alkalizing in the body despite their initial acidity. If we ate nothing but raw fruits, leafy greens and vegetables, our blood chemistry would be alkaline most of the time. The times when it would not be so would be in times of stress, or when we are exposed to environmental toxins, or are taking alcohol, caffeine or medication. Grains, most legumes, and most commonly eaten nuts are acid-forming (and mucus-forming) foods to some extent.

The chlorophyll in leafy green vegetables cleanses and alkalizes the blood. The body converts chlorophyll into heme, an iron compound that is part of hemoglobin, to produce red blood cells. Unfortunately, many Americans do not eat greens except in small

---

[100] https://www.jhsph.edu/news/news-releases/2016/diet%20-designed-to-lower-blood-pressure-also-reduces-risk%20of-kidney-disease.html.

amounts, such as lettuce in fast food sandwiches, which is one of the reasons why many people in this country are lacking in vitamins, antioxidants, and the therapeutic properties that plant foods have. Greens include spinach, kale, chard, lettuces, cabbages, collard greens and mustard greens. The importance of eating greens for health is discussed throughout this book.

Homeostasis is the tendency of the body to maintain itself in stable chemical equilibrium. Health is said to be a balancing act, with the body trying to balance or stabilize itself to a normal or alkaline blood condition. Obviously, the effectiveness at doing this is encumbered or enhanced by the foods that are eaten, and how they are eaten. An acidic blood condition is typically caused by a meat and dairy diet. If the diet is continued, the blood condition worsens to where the body's attempts at homeostasis are not sufficient in neutralizing the acidic condition. The acids the body cannot neutralize and expel as waste get stored in the tissues and joints of the body which can lead to diseases.

Two examples of the effects of a poor blood condition are anemia, a condition of not having enough healthy red blood cells, and deep-vein thrombosis, which is blood clotting. Healthy blood does not produce these disorders.

According to Dr. Michael Greger's book, *How Not to Die*, plant-based diets have been shown to reduce the risk of blood cancers by 50%.

pH Balance

Nature's way is for the human body to maintain its blood in an alkaline pH range of 7.35 - 7.45, the range that the body tries to maintain at all times through the process of homeostasis. pH is a term used in chemistry for the amount of acidity or alkalinity of an aqueous solution. The pH scale runs from 0 to 14, with a pH of 7 being neutral, a pH of less than 7 being acidic, and a pH greater than 7 being alkaline.

The basal pH of gastric juices secreted by the glands of the stomach is strongly acidic, with a range of 1.5 - 3.5. The pH in the stomach changes when food is in the stomach, and is influenced by a number of psychological factors, including the aroma and taste of foods.

Blood Sampling and Analysis

A blood test is often prescribed by doctors to help diagnose a person's health condition. The blood sample is sent to a laboratory where technicians analyze the blood using specialized instruments and techniques. Various tests may be performed on a blood sample, including a complete blood count (CBC), which is used to detect a wide range of health disorders such as anemia, infection and leukemia, and a blood glucose test which is used to help diagnose diabetes and monitor blood glucose levels. None of the tests are conclusive in themselves but are used to help diagnose a patient's condition and determine what follow-up tests should be prescribed.

In Appendix D of his book, *The Detox Miracle Sourcebook*, Robert Morse describes how anyone can interpret the results of the blood work that a doctor has prescribed for them. It includes a description of blood types (Types A, B, AB, etc.), the meaning of red and white blood cell counts, and the limitations that the diagnostics have. It lists reference ranges for nutrients that are found in the blood. The ranges are based on analyses of blood of presumably healthy people. But the point is, it seems that the body requires nutrients to be within certain ranges for health.

According to T. Colin Campbell in his book, *Whole, Rethinking the Science of Nutrition*. the body is continually monitoring and adjusting the concentrations of nutrients in the blood to maintain the ranges it requires for health. He explains how medical and governmental understanding of nutrition is rooted in the reductionist paradigm, a way of thinking that everything can be understood through its component parts. He contends that a wholistic approach to health is what is required to understand nutrition.

As explained in the chapter on the Causes of Disease, a wholistic approach considers how the various component parts of the body work together, which is how Nature operates. Nature works in wholistic ways, with all parts working together, never with one part working on its own.

"When you're looking through a microscope, either literally or metaphorically, you can't see the big picture." - T. Colin Campbell and Howard Jacobson, *Whole, Rethinking the Science of Nutrition*.

An old saying seems to be apropos here, "You can't see the forest for the trees." We cannot see the forest when we are focusing on the trees.

Nutritionists have known for many years that the condition of the urine reveals much about the blood's condition. For example, if the urine is cloudy, the blood is likely to be cloudy too, such as when protein intake has thickened it. The pH of urine closely matches the pH of the blood, and can be used to determine the blood's pH condition. Litmus paper is a useful tool in this regard. It is another example of how we can become more our own doctor. We can test our urine's pH. Strips of litmus paper may be purchased on the Web.

Chapter 6 The Way of Life

"Enter by the narrow gate; for wide is the gate and broad is the way that leads to destruction, and there are many who go in by it. Because narrow is the gate and difficult is the way which leads to life, and there are few who find it." - Jesus in the Bible, Matt. 7:13-14.

Many of the ties that exist between health disorders and poor nutrition were provided in the chapters on the Causes of Disease and More About Diet. But for most people, changing the situation by adopting a healthy diet counteracts the harm that they're doing to themselves.

Increasing scientific evidence compiled every year links the nominal top 10 leading causes of death, and the degenerative diseases so prevalent in the world today, to eating meat-based and dairy-based diets. These studies continue to show that people eating a plant-based diet have increased longevity and health compared to those eating a meat-based and/or dairy-based diet. Many books published in recent years provide the results of these clinical studies.

The China Study, published in 2005, may be the most comprehensive study of human nutrition ever performed. The population, or group, that was used in the study was the entire population of China. Written by T. Colin Campbell, a professor of Nutritional Biochemistry at Cornell University, and his son Thomas M. Campbell II, a physician, the study proved that whole plant-based foods, not animal-based foods, are the most beneficial foods for people. The study showed that people eating a plant-based diet have increased longevity and health compared to those eating a meat-based and/or dairy-based diet.

How Not to Die, written by Dr. Michael Greger and published in 2015, confirmed the conclusions of *The China Study*, and provided additional research and study results that emphasized the importance of eating plenty of whole plant foods, such as fruits and vegetables, to prevent and even *reverse* the chronic diseases of the Western world, including cancer, diabetes, heart disease and brain diseases.

The hazards of eating animal-based foods, or animal products, are well known to most educated people. Eating animal products causes plaque formations in the arteries, which is known to cause hardening of the arteries, which can lead to high blood pressure, heart disease and stroke. Harmful mutagens and carcinogens, such as acrylamide, HCAs, PAHs and AGEs, are formed when animal products are cooked (for more detail, see Victoria Boutenko's book, *12 Steps to Raw Foods*). Animal products can contain nitrates, chlorine and ammonia and are susceptible to hosting various forms of life-destroying bacteria.

Dr. Caldwell Esselstyn, Jr., in his book, *Prevent and Reverse Heart Disease,* says that anyone with high blood cholesterol levels is prone to heart disease. Animal-based foods contain cholesterol, whereas no plant foods contain cholesterol. Dr. Esselstyn changed his diet to a plant-based diet and strongly recommended his patients to do the same. Those who did were able to cleanse their coronary arteries of plaque formations, which means their arteries were no longer clogged, and Dr. Esselstyn proved this by way of coronary angiograms.

Today's real need is not another low-carb, high-protein diet, or an end to global warming, which appears to be mainly caused by extensive deforestation efforts to clear space for cattle and feed crops. Rather, it is self-education about foods and nutrition.

Typically, people resort to doctors when they don't know what else to do. But, by simply utilizing online sources and obtaining information that is contained in books that are available to most people, many health concerns can be thoroughly investigated and proper treatments determined, without seeing a physician.

How to Be Your Own Doctor

Several chapters have stressed the importance of diet change for health because of the strong ties that exist between the foods we eat and the body's natural ability to heal and ward off diseases. Emphasis has also been placed on the importance of taking charge of your health. More is said about this strategic breakthrough in this chapter.

No matter what age, gender or ethnic group, when a person starts eating healthy foods and learns how to properly combine foods, a doorway to health opens to them that may never shut again.

Numerous medical studies have attested to the fact that whole plant foods provide miraculous health benefits, and many books have been written about the incredible powers of raw plant foods to enable the body to cleanse itself of toxins and heal itself of diseases.

As you enter through the doorway to optimum health, you will find a marvelous and adventurous land verdant with resplendent gardens, orchards and colorful landscapes. The horizon stretches farther than the eye can see, and every day it rains at least once amidst plenty of sunshine.

As you take the exciting and rewarding journey into new realms of self-knowledge and self-awareness, you find that it enables you to become a more energetic and effective person as a knowledge of foods and nutrition equips you to become your own doctor using the untainted and unaltered foods of Nature. This means fewer doctor bills and less time spent in the doctor's office.

For those who resort to the care of doctors or other health care professionals for whatever ails them, it is important to get used to the idea that you are, not just in the final analysis but now and during each and every day of your life, responsible for your own health. Since your body is the only one that you will ever have in this world, it only makes sense that health should be a top priority in your life.

It is our responsibility to take care of our bodies. It is not our doctor's, our spouse's, our friends' or the Government's. It is our responsibility. We should avoid foods that are harmful to health and eat foods that promote health and longevity. These are whole plant foods, replete with their life-giving properties. These are the foods that God and Nature intended for us to eat.

"But the foods which you eat from the abundant table of God give you strength and youth to your body, and you will never see disease. For the table of God fed Methuselah of old, and I tell you truly, if you live as he lived, then will the God of the living give you also long life upon the earth as was his." - Attributed to Jesus, *The Essene Gospel of Peace, Book One.*

The following prayers are taken from *The Essene Gospel of Peace, Book One.*[101] The first is the well-recognized Lord's prayer, and the second the less familiar Earthly Mother prayer:

"After this manner, therefore, pray to your Heavenly Father: Our

---

[101]Edmond Bordeaux Szekely (Translator), The Essene Gospel of Peace, Book One, 1981

Father which art in heaven, hallowed be thy name. Thy Kingdom come. Thy will be done on earth as it is in heaven. Give us this day our daily bread. And forgive us our debts, as we forgive our debtors. And lead us not into temptation, but deliver us from evil. For thine is the kingdom, the power, and the glory, forever. Amen."

"And after this manner pray to your Earthly Mother: Our Mother which art upon earth, hallowed be thy name. Thy kingdom come, and thy will be done in us, as it is in thee. As thy sendest every day thy Angels, send them to us also. Forgive us our sins, as we atone all our sins against thee. And lead us not into sickness, but deliver us from all evil, for thine is the earth, the body, and the health. Amen."

Notice that the Lord's prayer asks for deliverance from temptation, and the Earthly Mother's prayer asks for deliverance from sickness. The connection between sickness and the foods we eat is so well established that those who pray the Lord's prayer could say, "And lead us not into the temptation to eat foods that are harmful to us, but deliver us from sickness and disease, for thine is the kingdom, the power, and the glory, forever. Amen."

Much of the joy, excitement and contentment we get out of life depends on how we take care of ourselves.

The main criterion for food choices for most people is how foods taste and go down rather than how they affect health. In other words, taste-appeal over health-appeal. Unfortunately, this truth has been exploited by the many concerns of the food industry in the creation of many substances that are added to our foods and drinks to make them more appetizing. The second basis for food selection is the emotional pull that certain foods have on us.

These criteria drive the majority of our decisions about foods. They also cause food habits to form.

Food habits are long-standing patterns of behavior associated with eating. Those of us who grew up on burgers, French fries and soft drinks tend to stick with that menu, or slight variations of it, throughout life, unless of course change is initiated.

The quantity of food consumed during meals also matters for health. It should not exceed a person's nutritional needs, for digestive juices are secreted not in proportion to the amount of food eaten, but in proportion to the amount of food that is required by the system.

"The major characteristic of the diet of longevous people is low total calorie intake throughout life." - Dan Georgakas, *The Methuselah Factors*.

The body always lets us know how we are treating it. Understanding and heeding the warning signals the body provides helps us to maintain ourselves in concert with the laws of Nature.

A basic truth about our lives is that the consequences of our choices are often different than their immediate effects. For example, to avoid the perils of constipation, one must not eat foods that cause it, notwithstanding the joy one may have in eating such foods.

We must examine our propensity to eat foods that are dictated by cultural norms and traditions, and foods that are advertised and otherwise promoted by the many interests of the food industry. We must examine why we depend on popular opinion to tell us what foods and drinks we should ingest. We must maintain a questioning attitude, and be more consumer-wise by reading food labels and knowing what they mean.

One would think that the mind would tell us exactly what to eat for mind and body health. But as everyone knows, that is not the case, for the mind is subtly influenced by many things, such as memories of past enjoyed meals and the price that was paid for them. As a result, most of the time our food choices are based on what we are most accustomed to eat, what is most available and convenient, and what is in keeping with cultural norms and traditions. In fact, unless the mind is sufficiently trained in the truths about foods and nutrition, it is a rare exception indeed that foods that have health-appeal over than taste-appeal are ever considered.

The foods we put into our bodies every day play a much larger role in determining our health and well-being than many people realize. Some nutritionists say that it is the most important determinant of health. All other things, including exercise, fresh air, sunshine and rest and relaxation are of lesser importance. However, moderate exercise and getting out of cramped quarters is very important, especially for those who lead sedentary lives.

As stated in the chapter on the Causes of Disease:

"Healing is no accident. All nature heals itself when causes are removed and the conditions of health supplied." - Dr. Herbert M. Shelton.

When animals get sick they instinctively abstain from all food. Can this be said about us? To a great extent, it cannot. Most people keep eating when they are sick on the mistaken belief that nourishment is needed for them to get well. Only rarely, such as in cases of acute fever, do we lose our appetite for food. Most doctors prescribe nourishment for those who are sick in hospitals, or who are otherwise under their care, and many patients force themselves to eat when the wisest thing to do would be to abstain from all food. It appears that in this context, animals are wiser than humans.

The healing powers of the body are fully at work when we live in concert with the laws of Nature. But when we depart from Nature's laws by putting foods in us that do us harm, or do other things that are bad for health, such as depriving the body of adequate exercise, sunshine, fresh air and rest, we court disaster in the form of disease or some other troublesome health issue. However, if we understand the laws of Nature and abide by them, illness and disease are vanquished by the body's inherent life promoting and sustaining powers.

"Healthy food gives us the energy to be healthy and happy. When we eat food with energy, we become people with energy. The difference between people with energy and people without energy is quire dramatic." - Sergei and Valya Boutenko, *Eating Without Heating*.

It was by learning about foods and nutrition that I was able to improve my health after suffering from many health issues for many years. I practice what I preach, and I believe that learning the truth versus hype about foods and nutrition is the best way to reach health goals. The knowledge gained during the learning process is priceless, for very few things are as important as good health.

Nobody is born with knowledge; it must be gained. When the knowledge required to promote and sustain health is gained and put into practice, practically any health issue that assails us can be effectively dealt with without the doctors' advice. It equips us to become, in many ways, our own doctors. This reality has far-reaching consequences for our health and well-being, providing numerous health benefits and preventing numerous health disorders.

"The excellence of knowledge is that wisdom gives life to those who have it." - Eccles. 7:12.

Some people like to be doctored, but it has never appealed to me unless I am truly ill. Because of the knowledge of foods and nutrition gained over the years, and having witnessed the healing powers of whole plant foods, I have learned that not only do I not need a doctor's advice on every little thing that may go wrong with me, but I can save myself a lot of time and tons of money by healing my own health issues, and even more time and money by not getting many of diseases common to man.

Another reason for avoiding doctors is the widespread prevalence of medical malpractice cases due to clinical errors made by practicing physicians and their aids.[102] As stated previously in the chapter on The Causes of Disease, these things have led many people to seek second or even third opinions about their health issues.

No one really knows the state of their health better than each of us ourselves. And whether people are aware of it or not, most are seeking a healthier, less stressful, more simple, peaceful and harmonious life.
But I'm glad that doctors are there when we really need them, like

---

[102] NCBI defines medical malpractice as any act or omission by a physician during treatment of a patient that deviates from accepted norms of practice in the medical community and causes injury to the patient.

for accidents and traumatic injury, or when we have reached the end of our rope and nothing that we do works to correct the issue. Being your own doctor has advantages besides saving money on hospital stays and doctor bills. The sense of accomplishment and the boost in self-esteem that follows being able to heal yourself of a health issue are priceless rewards. Is there a doctor in the house? Yes, and it's you!

Being your own doctor in essence means learning all you can about foods and nutrition from informed sources, such as the books that are referenced in the Bibliography, putting into practice what you learn from these sources, and heeding the signals the body gives about what should and should not be eaten. Sooner or later you realize that the symptoms of any sickness are signs aimed at getting you to change something about what you are doing.

All the theory in the world will not do you any good if you do not put theory into practice. Being your own doctor should be considered an on-going learning process putting into practice what is learned about foods and nutrition and other key principles of a healthy life.

"Life's greatest achievement is the continual remaking of yourself so that at last you know how to live." - Winfred Rhodes.

Next Steps

The book has shown that we can no longer choose to ignore information that has been gained in the human health field which proves that diseases and a host of lesser human ailments are primarily caused by commonly consumed foods, and that these same foods put us at much greater risk for being infected by viruses such as SARS-CoV-2 and its variants. For those who are vaccinated and those who are not, it is extremely important to ensure that your immune system is in good working order and being continually strengthened by the foods that you eat.

For additional information, I encourage you to read the books that are listed in the Bibliography. They have much to offer the novice as well as the long-time student of foods and nutrition. We all like to eat, but it is only when we learn to eat foods that promote health, not destroy it, that our health flourishes. The books provide the information and inspiration that will help all seekers of health. They have been a constant source of encouragement to me on my health journey. Take the time to delve into them and you will be glad that you did.

Bibliography

The following books were the major resources used to write this book.

1. Thomas S. Cowan MD, and Sally Fallon Morell, The Contagion Myth, 2020.

2. Joseph Mercola and Ronnie Cummins, The Truth About COVID-19, 2021.

3. Benjamin Franklin, Poor Richard's Almanacs, 1964.

4. Professor Arnold Ehret, Thus Speaketh the Stomach and the Tragedy of Nutrition, Introduced & Edited by Prof. Spira, 2014.

5. Horace Fletcher, A.M., Fletcherism: What It Is or How I Became Young at Sixty, 1913.

6. Carol Tavris, Anger, the Misunderstood Emotion, 1989.

7. Meyer Friedman, M.D. and Diane Ulmer, R.N., M.S., Treating Type A Behavior – and Your Heart, 1984.

8. Robyn Openshaw, The Green Smoothies Diet. 2021.

9. Dr. Joel Fuhrman, Eat to Live: The Amazing Nutrient Rich Program for Fast and Sustained Weight Loss, 2005.

10. Harvey Diamond, Fit for Life Not Fat for Life, 2003.

11. Harvey Diamond, Living Without Pain, 2007.

12. Robert O. Young and Shelly R. Young, The pH Miracle, 2010.

13. Paul C. and Patricia Bragg, The Miracle of Fasting, 2005

14. Dr. Edward Howell, Enzyme Nutrition, 1985.

15. Paul C. and Patricia Bragg, Water, The Shocking Truth That Can Save Your Life, 2004.

16. T. Colin Campbell, The China Study, 2006.

17. Dr. Michael Greger, How Not to Die, 2015.

18. John Smith, Fruits and Farinacea- The Proper Food of Man, 2015.

19. Russell T. Trall, Scientific Basis of Vegetarianism, 1970.

20. Dr. Caldwell Esselstyn, Jr., Prevent and Reverse Heart Disease, 2007.

21. Dr. Ann Wigmore, Be Your Own Doctor, 1982.

22. Dr. Ann Wigmore, Why You Do Not Have to Grow Old, 1985.

23. Arnold Ehret, The Mucusless Diet Healing System, 2015.

24. Arnold Ehret, Physical Fitness Through a Superior Diet, Fasting, and Dietetics , 2018.

25. Arnold Ehret, Rational Fasting and Roads to Health and Happiness, 2002.

26. Arnold Ehret, The Cause and Cure of Human Illness, 2001.

27. Teresa Mitchell, My Road to Health, 1987.

28. Norman W. Walker, Colon Health, 2005.

29. Norman W. Walker, Become Younger, 1978.

30. Norman W. Walker, Fresh Vegetable and Fruit Juices, 1978.

31. Norman W. Walker, Diet and Salad Suggestions, 1985.

32. Norman W. Walker, Water Can Undermine Your Health, 1995.

33. Victoria Boutenko, Green for Life, 2005.

34. Victoria Boutenko, 12 Steps to Raw Foods, 2005.

35. David Wolfe, The Sunfood Diet Success System, 2008.

36. Professor Spira, Spira Speaks, Dialogs and Essays on The Mucusless Diet Healing System, 2014.

37. Dr. Bernard Jensen, Guide to Diet and Detoxification, 2000.

38. Dr. Bernard Jensen, The Healing Power of Chlorophyll, 1973.

39. Karyn Calabrese, Soak Your Nuts, 2011.

40. Dr. Herbert M. Shelton, Superior Nutrition, 1994.

41. Dr. Herbert M. Shelton, Fasting Can Save Your Life, 1978.

42. Dr. Herbert M. Shelton, Food Combining Made Easy, 1982.

43. Dr. Russel Blaylock, Excitotoxins, The Taste that Kills, 1997.

44. Joe Alexander, Blatant Raw Foodist Propaganda, 2005.

45. Steve Meyerowitz, Sprouts, The Miracle Food, 1997.

46. Dr. Henry Lindlahr, Philosophy of Natural Therapeutics, 1975.

47. Dan Georgakas, The Methuselah Factors, 1980.

48. Alexander Leaf, M.D., Youth in Old Age, 1975.

49. Andrew Weil, M.D., Healthy Aging, 2005.

50. Robert Morse, N.D., The Detox Miracle Sourcebook, 2004.

51. Arnold Paul De Vries, Therapeutic Fasting, 1958

52. Dr. Kristine Nolfi, M.D., The Miracle of Living Foods, 1981.

53. T. Colin Campbell and Howard Jacobson, Whole: Rethinking the Science of Nutrition, 2014.

54. Wallace D. Wattles, Health Through New Thought and Fasting, 2007.

55. Francoise Wilhelmi de Toledo, MD, and Hubert Hohler, Therapeutic Fasting: The Buchinger Amplius Method, 2018.

56. Edward Hooker Dewey, M.D., The No-Breakfast Plan and the Fasting Cure. 1900.

57. Edward Hooker Dewey, M.D., The True Science of Living. 1894.

58. Upton Sinclair, The Fasting Cure, 1911.

59. Hereward Carrington, Vitality, Fasting and Nutrition, 1908.

60. Dr. Herbert M. Shelton, Health for the Millions, 1968.

61. O. L. M. Abramowski, Fruitarian Diet and Physical Rejuvenation, 1916.

62. Prof. Arnold Ehret's Mucusless Diet Healing System: Annotated, Revised, and Edited by Prof. Spira, 2015.

63. Sergei and Valya Boutenko, Eating Without Heating, 2002.

64. The Natural Hygiene Handbook, 1996.

65. NIH (National Institutes of Health) Record, April 2015.

66. G. Edmond Griffin, World Without Cancer, 2004.

67. Rich Anderson, Cleanse & Purify Thyself, 2000.

68. Howard F. Lyman, Mad Cowboy, 1998.

69. Luigi Cornaro, Sure Methods of Attaining a Long and Healthful Life, 1660.

70. Luigi Cornaro, The Surest Method of Correcting an Infirm Constitution, 1660.

71. Luigi Cornaro, How to Live 100 Years, or Discourses on the Sober Life, 1660.

72. Gabriel Cousens, M.D., Conscious Eating, 2000.

73. Jeffery M. Smith, Genetic Roulette, 2007.

74. F. Batmanghelidj, M.D, Your Body's Many Cries for Water, 2008.

75. Dr. Allen E. Banik, The Choice is Clear, 1989.

76. Eckhart Tolle, The Power of Now, 1999.

77. Tom Moore and Irene Pritikin, Pritikin: The Man Who Healed America's Heart, 1987.

78. Sherry A. Rodgers, M.D., The High Blood Pressure Hoax, 2005.

79. Robert E. Kowalski, The Blood Pressure Cure, 2007.

80. Robert Rowan, M.D, Control High Blood Pressure Without Drugs, 2001.

81. Alfred Armand Montapert, The Supreme Philosophy of Man: The Laws of Life, 1977.

82. Edmond Bordeaux Szekely (Translator), The Essene Gospel of Peace, Book One, 1981.

83. Stan Shepherd, Raw Veganism, 2018.

84. Stan Shepherd, Stop Sciatica and Spinal Stenosis, 2019.

85. S.H. Shepherd, How to Completely Get Rid of Hemorrhoids Naturally: A Permanent Cure, 2019.

86. Dr. Dean Ornish, Reversing Heart Disease, 1992.

87. S. H. Shepherd, A Christian Diet, 2019.

88. S.H. Shepherd, The Cure for Arthritis, 2020.

89. Annie Payson Call, Power Through Repose, 1905.

90. Charles Spurgeon, Beside Still Waters, 1999.

About the Author

S. H. Shepherd, 73, has researched and studied the human health field for over 30 years. An engineer by training, his extensive knowledge of the human health field qualifies him to write this book. He has witnessed the decline of health in this country and the erosion of the quality of life that people suffer who have been stricken with various health issues because of the lack of vital nutrients, phytochemicals and enzymes due to commonly eaten foods. The importance of telling others about what he has learned about the many health hazards associated with commonly eaten foods and the risk they pose on getting more serious diseases, and the facts about the COVID-19 pandemic, which he and others have held a keen interest in from the start, has been the incentive for writing this book..

Appendix I The Dirty Dozen and The Clean Fifteen

Several years ago, the Environmental Working Group (EWG) published lists of fruits and vegetables known as the Dirty Dozen and the Clean Fifteen. These lists indicate foods with the most and least pesticide residues on them based on data compiled by the USDA. The lists are updated annually. They reflect pesticide residues found on foods after they were washed with water.

The Dirty Dozen

The foods highest on this list have the most pesticides on them.

1. Strawberries
2. Apples
3. Nectarines
4. Peaches
5. Celery
6. Grapes
7. Cherries
8. Spinach
9. Tomatoes
10. Sweet bell peppers
11. Cherry tomatoes
12. Cucumbers

According to EWG, buying organic for the twelve fruits and vegetables on this list can reduce our pesticide exposure by at least 90 percent!

The Clean Fifteen

The foods highest on this list have the least pesticides on them.

1. Avocados
2. Sweet corn
3. Pineapples
4. Cabbage
5. Sweet peas (frozen)
6. Onions
7. Asparagus
8. Mangoes
9. Papayas
10. Kiwi
11. Eggplant
12. Honeydew melon
13. Grapefruit
14. Cantaloupe
15. Cauliflower

There is no need to buy organic for the fruits and vegetables on this list, except for cabbage (number 4) and papayas (number 9). For cabbage, according to David Wolfe's book, The Sunfood Diet Success System, non-organic cabbage has large amounts of pesticides are used on it. Papayas are GMO foods and have pesticides on them.

Some types of produce are more prone to containing pesticides residues than others. Avocados, sweet corn and pineapples, for example, are not so prone because of their protective outer layer of skin. Not the same for strawberries and other berries.

Index

A

Acidity, 25, 55, 68, 76-77, 79.

AGEs (advanced glycation end-products), 58, 82.

Alcohol, 77.

Antioxidants, 29, 44, 58, 62, 64-65, 78.

B

Brain disease, 82.

C

Cancer, 28, 32-33, 35-36, 38, 43, 46, 54, 56, 58, 72-73, 78, 82.

CDC (Centers for Disease Prevention and Control), 9, 11-16, 21-22.

Constipation, 32, 34, 41, 44-45, 55, 85.

D

Detoxification, 37.

Diabetes, 29, 35-36, 58, 79, 82.

Diets

    American (traditional), 47, 50, 58, 71.

    Vegetarian, 60-62.

    Vegan, 60-62.

    Raw vegan, 52, 61-62, 68.

Distilled water, 61.

E

Enzymes, 29, 36, 44, 49, 56-57, 62-63, 65, 68, 75.

Made in the USA
Monee, IL
08 January 2023

24875750R00059